Tiger Economies
Under Threat

Tiger Economies Under Threat

A Comparative Analysis of Malaysia's Industrial Prospects and Policy Options

Shahid Yusuf
Kaoru Nabeshima

THE WORLD BANK
Washington, D.C.

© 2009 The International Bank for Reconstruction and Development / The World Bank
1818 H Street NW
Washington DC 20433
Telephone: 202-473-1000
Internet: www.worldbank.org
E-mail: feedback@worldbank.org

All rights reserved

1 2 3 4 12 11 10 09

This volume is a product of the staff of the International Bank for Reconstruction and Development / The World Bank. The findings, interpretations, and conclusions expressed in this volume do not necessarily reflect the views of the Executive Directors of The World Bank or the governments they represent.

The World Bank does not guarantee the accuracy of the data included in this work. The boundaries, colors, denominations, and other information shown on any map in this work do not imply any judgement on the part of The World Bank concerning the legal status of any territory or the endorsement or acceptance of such boundaries.

Rights and Permissions
The material in this publication is copyrighted. Copying and/or transmitting portions or all of this work without permission may be a violation of applicable law. The International Bank for Reconstruction and Development / The World Bank encourages dissemination of its work and will normally grant permission to reproduce portions of the work promptly.

For permission to photocopy or reprint any part of this work, please send a request with complete information to the Copyright Clearance Center Inc., 222 Rosewood Drive, Danvers, MA 01923, USA; telephone: 978-750-8400; fax: 978-750-4470; Internet: www.copyright.com.

All other queries on rights and licenses, including subsidiary rights, should be addressed to the Office of the Publisher, The World Bank, 1818 H Street NW, Washington, DC 20433, USA; fax: 202-522-2422; e-mail: pubrights@worldbank.org.

ISBN: 978-0-8213-7880-9
eISBN: 978-0-8213-8061-1
DOI: 10.1596/978-0-8213-7880-9

Library of Congress Cataloging in Publication Data has been applied for.

Cover design by Drew Fasick, Serif Design Group.

Contents

Acknowledgments xv
About the Authors.. xvii
Abbreviations xix

Chapter 1. Southeast Asia Faces Mounting Competition 1
Brief Development History of the Four Southeast Asian Tigers............ 4
The East Asian Model.. 5
The Malaysian Experience 7
Analyzing Industrial Change in Southeast Asia 14
Tiger in the Spotlight 15

Chapter 2. Malaysia: The Quintessential Maturing Tiger Economy 17
Sources of Growth.. 17
Evolution of the Manufacturing Industry in Malaysia 22
Annex 2.A: Indicators of Competitiveness 31

Chapter 3. Analyzing Comparative Advantage and Industrial Change: Reading the Export Trade Tea Leaves 37
Exports and Industrial Change 38
An Overview of Export Capabilities 86

Chapter 4. Imports and Foreign Direct Investment: Competition and Technology Transfer 93
Imports and Technology Transfer 95
Patterns of Foreign Direct Investment 98
Technology Infusion from FDI and Upgrading 102

Chapter 5. Leading and Faltering Industries: The Electronics, Auto Parts, and Agro-Processing Sectors.. 105
Electronics and Electrical Engineering 105
Auto Parts Industry.. 109
Palm Oil, Biodiesel, and Food Products 113

v

Chapter 6. Can Southeast Asian Tiger Economies Become Innovative? 119
Industrial Location 119
Quality of Labor 131
Access to Finance... 150

Chapter 7. From Technology Development to Innovation Capability 159
R&D Spending... 160
Patenting Activity 172
Licensing and Technology Transfer.................................... 178
Research Activities of Malaysian Firms................................ 180
Innovation Comparative Advantage 183

Chapter 8. Can the Tigers Grow Fast and Furious Again? 187
Long-Run Growth.. 187
How Neighboring Economies Can Affect Malaysia...................... 199

Chapter 9. What Can the Tigers Do? 203

Appendix A
Revealed Comparative Advantage of East Asian Economies Other
than Malaysia... 219

Appendix B
Product Space Analysis for Southeast Asian Economies 225

Appendix C
Research and Development Spending by Private Firms in Malaysia 235

Appendix D
Index of Innovation Revealed Comparative Advantage 237

Appendix E
Financial Incentives for Research and Development, Technology
Development, and Innovation in Chinese Firms......................... 241

Appendix F
Financial Incentives for Research and Development, Technology
Development, and Innovation in Thai Firms............................ 243

References 249

Index... 269

Figures
2.1   Industrial Composition by Type of Manufacture, Malaysia,
      1981, 1990, and 2002 23
2.2   Value-Added Ratios in Machinery in Selected East Asian Economies.... 27

3.1 Share of Malaysia's and Selected Southeast Asian Countries' Exports to the World, 1995–2007 40
3.2 Share of Malaysia's and Selected Southeast Asian Countries' Exports to the United States, 1995–2007 40
3.3 Share of Malaysia's and Selected Southeast Asian Countries' Exports to China, 1995–2007. 41
3.4 Exports of Malaysia by Type of Manufacture, 1995–2007 41
3.5 Exports of Malaysia by Type of Manufacture, Excluding Electronics, 1995–2007 42
3.6 Exports of Electronic and Electrical Manufactures, 1995–2007......... 43
3.7 Composition of Exports by Type of Manufactures, 1995 and 2007...... 46
3.8 Share of Overlapping Commodities, 1995, 2000, and 2007 48
3.9 Share of Overlapping Trade Values, 1995, 2000, and 2007 48
3.10 Imports of Electronic Components by China from East Asian Countries, 1995–2007. 61
3.11 Imports of Electronic Components by China from East Asian Countries, Excluding Japan and Korea, 1995–2007. 62
3.12 Imports of Electronic Components by China, According to Economy of Origin, 1995–2007. 63
3.13 Exports of Electronic Components by China According to Economy of Destination, 1995–2007 63
3.14 Unit Values of Digital Monolithic Integrated Circuits, 1995–2007 70
3.15 Unit Values of Nondigital Monolithic Integrated Circuits, 1995–2007 ... 71
3.16 Unit Values of Hybrid Integrated Circuits, 1995–2007 71
3.17 Unit Values of Electronic Integrated Circuits, 1995–2007.............. 72
3.18 Unit Values of Parts of Electronic Integrated Circuits and Similar Items, 1995–2007 72
3.19 Unit Values of Color Cathode Ray Television Picture Tubes and Similar Items, 1995–2007 73
3.20 Product Space of Selected Southeast Asian Countries, 2000–04 83
3.21 Product Space of China, 2000–04 85
3.22 Composition of Service Exports from Malaysia, 1995–2007 88
3.23 Composition of Selected Service Exports from Malaysia, 1999–2007.... 88
3.24 Service Exports from Southeast Asian Countries, 1995–2007 89
4.1 Average Tariffs on Total Imports, 1995–2007........................ 94
4.2 Average Tariffs on Machinery and Equipment Imports, 1995–2007 94
4.3 Composition of Machinery and Equipment Imports by Malaysia, 1995–2007 96
4.4 Net Inflows of FDI, 1995–2006 98
4.5 Top 12 FDI Sectors in Malaysia, 1999–2003......................... 99
4.6 Top 12 FDI Sectors in Malaysia, 1999–2003........................ 100
4.7 Composition of Radio, Television, and Communication FDI Inflows into Malaysia by Source Economy, 1999–2003................ 101

6.1	Secondary School Gross Enrollment Ratio, 2006.	135
6.2	Hybritech and Its Daughter Firms in San Diego	142
6.3	Information and Communication Technology Investment and Labor Productivity Growth, 1989–2005	148
6.4	M2 as a Percentage of GDP, 1995–2007	151
6.5	Domestic Credit to Private Sector, 1995–2007	152
6.6	Number of Deals in Malaysia, Singapore, and Thailand, Including All Stages, 1990–2007	154
7.1	R&D Expenditure as a Share of GDP, 1996–2004	161
7.2	Total R&D Personnel, 1999–2004	162
7.3	Number of Researchers per Million People, 1996–2004	163
7.4	Share of R&D Spending by Businesses, 2002–04	164
7.5	Distribution of R&D by Sectors, 2007	165
7.6	Number of Published Papers in Professional Journals	171
7.7	Malaysia's U.S. Patents Relative to Those of Other Countries, 1977–2006.	175
7.8	Royalty and License Fee Receipts, 1995–2006	179
7.9	Royalty and License Fee Payments, 1995–2006	179
7.10	Net Royalty and Licensing Payments, 1995–2006	180
8.1	Per Capita GDP Growth of Germany, 1860–2003	188
8.2	Per Capita GDP Growth of the United Kingdom, 1860–2003	188
8.3	Per Capita GDP Growth of the United States, 1860–2003	189
8.4	Per Capita GDP Growth of Indonesia, 1960–2003	190
8.5	Per Capita GDP Growth of Japan, 1960–2003	190
8.6	Per Capita GDP Growth of the Republic of Korea, 1960–2003	191
8.7	Per Capita GDP Growth of Malaysia, 1960–2003	191
8.8	Per Capita GDP Growth of the Philippines, 1960–2003	192
8.9	Per Capita GDP Growth of Taiwan, China, 1960–2003.	192
8.10	Per Capita GDP Growth of Thailand, 1960–2003	193
8.11	Average Years of Schooling, 1960–2000	194
8.12	Per Capita GDP Growth and Labor Force with Tertiary Education: Germany, 1860–2000	194
8.13	Per Capita GDP Growth and Labor Force with Tertiary Education: Japan, 1960–2000	195
8.14	Per Capita GDP Growth and Labor Force with Tertiary Education: Republic of Korea, 1960–2000	195
8.15	Per Capita GDP Growth and Labor Force with Tertiary Education: Taiwan, China: 1960–2000	196
8.16	Per Capita GDP Growth and Labor Force with Tertiary Education: United States, 1860–2000	196
8.17	Per Capita GDP Growth and Labor Force with Tertiary Education: Indonesia, 1960–2000	197

8.18 Per Capita GDP Growth and Labor Force with Tertiary Education: Malaysia, 1960–2000 197
8.19 Per Capita GDP Growth and Labor Force with Tertiary Education: Philippines, 1960–2000 198
8.20 Per Capita GDP Growth and Labor Force with Tertiary Education: Thailand, 1960–2000 198

Tables

1.1 Export Shares of Primary and Resource-Based Products 5
1.2 Contribution of Agriculture 5
1.3 Export Structure of Southeast Asian Countries, 1970–2007 9
2.1 Historical Performance of Malaysia 18
2.2 Average Sectoral Contribution to Growth: Demand Side 18
2.3 Average Sectoral Contribution to Growth: Supply Side 19
2.4 Contributions of Input Growth and TFP Growth 20
2.5 TFP Growth for Malaysia and Other Countries 21
2.6 Malaysia's Share in World Production, 1981, 1990, and 2002 24
2.7 Value Added per Worker in Malaysia and Its Growth Rates, 1981, 1990, and 2002 25
2.8 Average Wage in Malaysia and Its Growth Rates, 1981, 1990, and 2002 .. 26
2.9 Ratio of Value Added to Output Value, 1981, 1990, and 2002 27
2.10 Rank Correlation among Selected Indexes on Technological Capabilities 28
2.11 Macroindicators of Malaysia and Other Southeast Asian Economies, 1990–2007 29
2.A.1 IMD Global Competitiveness Ranking, 2004–08 31
2.A.2 Subcomponents to IMD Global Competitiveness Rankings: Malaysia, 2003–07 31
2.A.3 WEF Global Competitiveness Index, 2006–07 and 2007–08 32
2.A.4 WEF Global Competitiveness Rankings of Malaysia and Its Major Competitors, 2007–08 33
2.A.5 Doing Business Indicators, 2006–09 34
2.A.6 Individual Components of Doing Business Rankings: Malaysia, 2005–08 35
3.1 Exports and Imports of Goods and Services as a Percentage of GDP, 2007 38
3.2 Exports of Goods and Services, 2000–07 39
3.3 Fastest-Growing Exports from Malaysia by Level of Technology, 2000–07 44
3.4 Leading Export Sectors, 1995 and 2007 45
3.5 Share of Overlapping Commodities, 1995, 2000, and 2007 47
3.6 Share of Overlapping Trade Values, 1995, 2000, and 2007 47

3.7	Full Time Period, 1990–2006, with Subperiod Dummies	50
3.8	Degree of Competition between Malaysian and Chinese Exports, 1990–2007	52
3.9	Dynamic Revealed Competitiveness Position for Malaysia's Top 10 Nonoil Exports to the United States, 2007	52
3.10	Dynamic Revealed Competitiveness Position for Malaysia's Top 10 Nonoil Exports to Japan, 2007	53
3.11	Dynamic Revealed Competitiveness Position for Malaysia's Top 10 Exports to the EU15, 2007	54
3.12	Malaysia's Most Dynamic Export Goods to the United States, 1990–2007	55
3.13	Malaysia's Most Dynamic Export Goods to Japan, 1990–2007	56
3.14	Malaysia's Most Dynamic Export Goods to the EU15, 1990–2007	57
3.15	Dynamic Revealed Competitiveness Position by Technology Level in U.S. Market: Malaysia versus China, 1991–2007	58
3.16	Dynamic Revealed Competitiveness Position by Technology Level in Japanese Market: Malaysia versus China, 1991–2007	59
3.17	Dynamic Revealed Competitiveness Position by Technology Level in EU15's Market: Malaysia versus China, 1991–2007	59
3.18	Investments in Electrical and Electronics Industry, 1996–2005	64
3.19	Number of Commodities That Entered or Exited Selected Asian Export Baskets, by Category	64
3.20	Percentage of Commodities That Entered or Exited Selected Asian Export Baskets, by Category	65
3.21	Fastest-Growing "New" Exports from Malaysia, 2000–07	66
3.22	Top 10 Fastest-Growing Electronics and Electrical Exports from Malaysia, 2000–07	67
3.23	Top 10 Fastest-Growing Electronics and Electrical Exports from Thailand, 2000–07	68
3.24	Top 10 Fastest-Growing Electronics and Electrical Exports from the Philippines, 2000–07	68
3.25	Top 10 Fastest-Growing Electronics and Electrical Exports from China, 2000–07	69
3.26	New Electronic and Electricals That Appeared from 1995 to 2006 and Their 2000–04 PRODY	76
3.27	Destination of Malaysia's Exports, 1995, 2000, and 2007	78
3.28	Top 10 Commodities with Highest Revealed Comparative Advantage in Malaysia, 1995	79
3.29	Top 10 Commodities with Highest Revealed Comparative Advantage in Malaysia, 2000	79
3.30	Top 10 Commodities with Highest Revealed Comparative Advantage in Malaysia, 2007	80

3.31	Top 10 Malaysian Exports in 2007 and Their Revealed Comparative Advantages in 1995–2007	81
3.32	Costs of Selected Procedures in Selected Countries	90
5.1	Contribution of the Electronics and Electrical Machinery, Automotive, and Palm Oil Industries to Manufacturing GDP, 2005	106
6.1	City Population and Share of National Population, 2005	123
6.2	Cities' Share of National GDP, 1985–2005	124
6.3	Population and Its Growth Rates in Four Cities in Malaysia	125
6.4	GDP and Its Growth Rates in Four Cities in Malaysia	125
6.5	FDI and Its Growth Rates in Four Cities in Malaysia	126
6.6	Top 10 Net New Entrants in Manufacturing: Kuala Lumpur, 1995–2006	127
6.7	Top 10 Net New Entrants in Manufacturing: Johor, 1995–2006	128
6.8	Top 10 Net New Entrants in Manufacturing: Penang, 1995–2006	129
6.9	Top 10 Net New Entrants in Manufacturing: Malacca, 1995–2006	130
6.10	Distribution of Labor Force by Educational Attainment, 2002 and 2006	131
6.11	Share of Firms That Reported Vacancies for Various Occupations, 2007	131
6.12	Time It Takes to Fill the Position, 2007	132
6.13	Most Critical Skill in Shortage, 2002 and 2007	132
6.14	Most Important Causes of Vacancy	133
6.15	Eighth-Grade TIMSS Scores for Mathematics for Selected East Asian Economies, 1999, 2003, and 2007	135
6.16	Eighth-Grade TIMSS Scores for Science for Selected East Asian Economies, 1999, 2003, and 2007	136
6.17	Ranking of Selected Universities in East Asia, 2007 and 2008	137
6.18	Number of Government Research Institutes, 2008	138
6.19	Location of Public and Private Universities, 2008	139
6.20	Malaysia's 40 Largest Firms with Spinoff Potential, 2007	143
6.21	Locations of Headquarters	144
6.22	Broadband Subscribers, 2001–07	145
6.23	Information Technology and Innovation Foundation Broadband Rankings, 2006	146
6.24	Average Speed of Internet Connections Offered in Malaysia, 2008	146
6.25	International Bandwidth, 1999–2005	147
6.26	E-Readiness Scores, 2003–07	149
6.27	Volume of Venture Capital from Public and Private Sources in Malaysia, 2003–06	153
6.28	Cumulative Deals in Malaysia, by Sector, 1990–2007	155
6.29	Initial Public Offerings by Malaysian Firms	155
7.1	R&D Spending as a Share of GDP, 1996–2006	160

7.2	Total R&D Personnel Nationwide, 1997–2004	161
7.3	Top Five Sectors with Highest R&D Spending in Malaysia, 2006	164
7.4	R&D Spending by Major Automobile Firms, 2007–08	166
7.5	R&D Spending by Top Four GRIs, 2004	166
7.6	Top 15 Fields of Emphasis by GRIs, 2004	167
7.7	Distribution of R&D Researchers in GRIs with More Than 100 Researchers, 2004	168
7.8	R&D Spending by Top Research Universities, 2004	169
7.9	Top 15 Fields of Emphasis by Universities, 2004	169
7.10	Research Concentration by Fields, 2004	170
7.11	Top 10 R&D Spending by GRIs and Universities, 2004	171
7.12	Patents Granted in Malaysia, 1995–2007	172
7.13	Top Holders of Malaysian Patents: Private Firms, 1989–2006	173
7.14	Top 10 Patent Owners in Malaysia, 1995–2007	173
7.15	Distribution of Domestic Patents by Technology Class, 2000–07	174
7.16	Patents Granted by U.S. Patent and Trademark Office to Foreign Residents	174
7.17	U.S. Patents Granted to Malaysian Organizations, 2003–07	176
7.18	U.S. Patent Classes Granted to Malaysian Residents, 2003–07	177
7.19	Number of U.S. Patents Granted in the Active Solid-State Devices Technology Class, by Geographic Origin, 2003–07	177
7.20	Number of U.S. Patents Granted in Semiconductor Device Manufacturing, Process Technology Class, by Geographic Origin, 2003–07	178
7.21	Index of Innovation Revealed Comparative Advantage, by Technology Class: Malaysia, 1995	184
7.22	Index of Innovation Revealed Comparative Advantage, Top 10 Technology Classes: Malaysia, 2000	184
7.23	Index of Innovation Revealed Comparative Advantage, Top 10 Technology Classes: Malaysia, 2007	185
7.24	Index of Innovation Comparative Advantage, Top 10 Technology Classes: Singapore, 2007	185
8.1	Average Growth Rates of GDP Per Capita, 1870–2003 and 1960–2003	189
8.2	R&D Expenditure, 1985–2006	193
9.1	Malaysia's Incentive Policies	204
9.2	List of Major Investment Incentives Available for the Manufacturing Sector	206
9.3	Malaysia: Highest IRCA in 1995 and RCA in 2000	210
9.4	Malaysia: Highest IRCA in 2000 and RCA in 2005	211
9.5	Thailand: Highest IRCA in 1995 and RCA in 2000	212
9.6	Thailand: Highest IRCA in 2000 and RCA in 2005	213

A.1	Top 10 Commodities with Highest Revealed Comparative Advantage in China, 2007	219
A.2	Top 10 Commodities with Highest Revealed Comparative Advantage in Indonesia, 2007	220
A.3	Top 10 Commodities with Highest Revealed Comparative Advantage in Japan, 2007	220
A.4	Top 10 Commodities with Highest Revealed Comparative Advantage in the Republic of Korea, 2007	221
A.5	Top 10 Commodities with Highest Revealed Comparative Advantage in the Philippines, 2007	221
A.6	Top 10 Commodities with Highest Revealed Comparative Advantage in Singapore, 2007	222
A.7	Top 10 Commodities with Highest Revealed Comparative Advantage in Taiwan, China, 2007	222
A.8	Top 10 Commodities with Highest Revealed Comparative Advantage in Thailand, 2007	223
B.1	Top 20 Upscale Commodities with Highest Density in China, 2000–04	225
B.2	Top 20 Upscale Commodities with Highest Density in Hong Kong, China, 2000–04	226
B.3	Top 20 Upscale Commodities with Highest Density in India, 2000–04	227
B.4	Top 20 Upscale Commodities with Highest Density in Indonesia, 2000–04	228
B.5	Top 20 Upscale Commodities with Highest Density in the Republic of Korea, 2000–04	229
B.6	Top 20 Upscale Commodities with Highest Density in Malaysia, 2000–04	230
B.7	Top 20 Upscale Commodities with Highest Density in the Philippines, 2000–04	231
B.8	Top 20 Upscale Commodities with Highest Density in Singapore, 2000–04	232
B.9	Top 20 Upscale Commodities with Highest Density in Thailand, 2000–04	233
C.1	Breakdown of Private Research and Development Spending	235
D.1	Innovation Revealed Comparative Advantage: Top 10 Technology Classes in China, 2007	237
D.2	Innovation Revealed Comparative Advantage: Top 10 Technology Classes in India, 2007	237
D.3	Innovation Revealed Comparative Advantage: Top 10 Technology Classes in Indonesia, 2007	238
D.4	Innovation Revealed Comparative Advantage: Top 10 Technology Classes in the Philippines, 2007	238

D.5	Innovation Revealed Comparative Advantage: Top 10 Technology Classes in Singapore, 2007	239
D.6	Innovation Revealed Comparative Advantage: Top 10 Technology Classes in Thailand, 2007	239
E.1	Financial Incentives for Innovation Offered in China.	241
F.1	Support for Investments in the Development of Skills, Technology, and Innovation in Thailand	244
F.2	Expanded Support Programs for Enhancing Technology in Industry in Thailand	246
F.3	Support under the Revised Policy on Intellectual Property in Thailand	248

Acknowledgments

This volume is one of a series of publications emerging from a project on East Asia's Prospects, cosponsored by the government of Japan. It was done in close collaboration with the East Asia and Pacific Regional Office in the World Bank and the Knowledge Economy Section of the Economic Planning Unit of the Prime Minister's Department. The study benefitted from interaction with numerous firms, university personnel, and state-level authorities and institutions in Malaysia. We would like to express our deep appreciation to Tan Sri Dr. Sulaiman Mahbob, Dato' Noriyah Ahmad, and Mr. K. Yogeesvaran for their support and encouragement. We gratefully acknowledge the financial backing of the government of Japan through its Policy and Human Resources Development Fund for the research and the publication of the book. For their highly constructive comments and suggestions, we thank Milan Brahmbhatt, Mark Dutz, Bruno Laporte, Vikram Nehru, Ian Porter, Omporn Regel, Tatyana Soubbotina, Mathew Verghis, Al Watkins, and Albert Zeufack. Lopamudra Chakraborti provided excellent research support throughout, and the expert assistance provided by Rebecca Sugui, Paulina Sintim-Aboagye, Audrey Kitson-Walters, and Imran Hafiz enormously facilitated the production of this volume. As in the past, we relied heavily on Patricia Katayama, Cindy Fisher, and Nora Ridolfi at the World Bank Office of the Publisher to manage the operation from start to finish, and the editors at Publications Professionals once again did a fine job of editing the manuscript. Our debt to them grows longer with each succeeding publication.

About the Authors

Shahid Yusuf is economic adviser in the World Bank Institute of the World Bank and currently manages a major project on East Asia's Prospects. He was the director for the *World Development Report 1999/2000, Entering the 21st Century* and has held positions in the Bank's regional and research departments. He received his B.A. in economics from Cambridge University and his Ph.D. in economics from Harvard University.

He has written extensively on development issues, with a special focus on East Asia. His most recent publications include *Postindustrial East Asian Cities*, co-authored with Kaoru Nabeshima (2006); *Dancing with Giants*, co-edited with L. Alan Winters (2007); *How Universities Promote Economic Growth*, co-edited with Kaoru Nabeshima (2007); *China Urbanizes*, co-edited with Tony Saich (2008); *Growing Industrial Clusters in Asia*, co-edited with Kaoru Nabeshima and Shoichi Yamashita (2008); *Accelerating Catch-Up: Tertiary Education and Growth in Africa*, co-authored with William Saint and Kaoru Nabeshima (2008); and *Development Economics through the Decades* (World Bank 2009).

Kaoru Nabeshima is a consultant for the World Bank Institute of the World Bank and previously served as an economist in the Bank's Development Research Group. He received his B.A. in economics from Ohio Wesleyan University. He holds a Ph.D. in economics from the University of California, Davis.

His recent publications include *Postindustrial East Asian Cities*, co-authored with Shahid Yusuf (2006); *How Universities Promote Economic Growth*, co-edited with Shahid Yusuf (2007); *Growing Industrial Clusters in Asia*, co-edited with Shahid Yusuf and Shoichi Yamashita (2008); and *Accelerating Catch-Up: Tertiary Education and Growth in Africa*, co-authored with Shahid Yusuf and William Saint (2008).

Abbreviations

AMD	Advanced Micro Devices
ASEAN	Association of Southeast Asian Nations
BIOS	basic input-output system
CCL	Cambridge Consultants
DDIT	double-deduction incentive for training
DRC	dynamic revealed competitiveness
DRCP	dynamic revealed competitiveness position
ETDZ	Economic and Technological Development Zone (China)
EU	European Union
FDI	foreign direct investment
GCC	government-controlled corporation
GDP	gross domestic product
GRI	government research institute
GTAP	Global Trade Analysis Project
HRDF	Human Resource Development Fund (Malaysia)
HS	harmonized system
HT1	electronic and electrical products
HT2	other high-tech products
HTIZ	high-tech industrial zone
ICT	information and communication technology
IMD	International Institute for Management Development
IPO	initial public offering
IRCA	innovation revealed comparative advantage
IT	information technology
ITA	investment tax allowance
JCI	Joint Commission International
LED	light-emitting diode
LT1	textiles, garments, and footwear
LT2	other low-technology products

MARDI	Malaysian Agricultural Research and Development Institute
MAVCAP	Malaysia Venture Capital Management
MESDAQ	Malaysian Exchange of Securities Dealing and Automated Trading
MIDA	Malaysian Industrial Development Authority
MIMOS	Malaysian Institute of Microelectronic Systems
MNC	multinational corporation
MPOB	Malaysian Palm Oil Board
MSC	Multimedia Super Corridor (Malaysia)
MT1	automotive products
MT2	process industry
MT3	engineering products
NASDAQ	National Association of Securities Dealers and Automated Quotations
NCER	Northern Corridor Economic Region (Malaysia)
NEP	New Economic Policy (Malaysia)
NSTDA	National Science and Technology Development Agency (Thailand)
OECD	Organisation for Economic Co-operation and Development
OLS	ordinary least squares
OSP2	Second Outsource Partners Programme (Malaysia)
PP	primary products
PSDC	Penang Skills Development Centre (Malaysia)
RA	reinvestment allowance
RB1	agro-based products
RB2	other resource-based products
R&D	research and development
RCA	revealed comparative advantage
SAIC	Shanghai Automotive Industry Corporation
SIRIM	Standards and Industrial Research Institute of Malaysia
SITC2	Standard Industrial Trade Classification revision 2
SMEs	small and medium-size enterprises
S&T	science and technology
STEM	science, technology, engineering, and math
STEP2	Second Science and Technology Policy (Malaysia)
TEFT	Technology Transfer from Research Institutes to SMEs (program) (Norway)
TFP	total factor productivity
TIMSS	Trends in International Mathematics and Science Study
UKM	Universiti Kebangsaan Malaysia

UM	University of Malaya
UN Comtrade	United Nations Commodity Trade Statistics Database
UNDP	United Nations Development Programme
UNIDO	United Nations Industrial Development Organization
UPM	Universiti Putra Malaysia
USM	Universiti Sains Malaysia
USPTO	U.S. Patent and Trademark Office
UTM	Universiti Teknologi Malaysia
WEF	World Economic Forum

1

Southeast Asia Faces Mounting Competition

Japan showed the way. Its remarkable economic recovery starting in the early 1950s, and sustained rapid growth through the early 1970s served to define the iconic East Asian industrialization cum export-led model of growth. Four economies followed in Japan's wake: Hong Kong, China; the Republic of Korea; Singapore; and Taiwan, China. All of these economies consciously emulated elements of the Japanese approach appropriate for their size, their history, and the prevailing international circumstances.[1] All hitched their future performance to industrialization, as Japan had done, with light manufacturing as the initial stepping-stone; all relied largely on rising rates of domestic savings to finance development; and all came to depend on domestic investment and export demand, mainly from the United States, to buttress their growth. Unlike Japan, however, Hong Kong, China; Singapore; and, to a lesser extent, Taiwan, China, fueled their early industrialization with the help of foreign direct investment (FDI), the bulk of it by companies from the United States. FDI also helped to hook firms in these economies to buyer-driven international value chains, which germinated in the 1960s as major U.S. retailers began sourcing more labor-intensive products from overseas and diversified away from Japanese suppliers.[2] Of the four, Korea relied the least on FDI (see Chung 2007). Home-grown companies, several of which had struck root in the first half of the 20th century, spearheaded Korea's industrial development and trade as they did in Japan (see Chung 2007; Graham 2003; Lie 2000).

The governments of Korea; Singapore; and Taiwan, China, adopted the Japanese approach of targeting and grooming potential leading sectors with the

---

[1]This concept has been captured by the flying geese model, in which Japan is depicted as the goose in the lead.

[2]The literature on these chains and how they operate is now vast. See Feenstra and Hamilton (2006); Gereffi and Korzeniewicz (1993); Schmitz (2005); and Yusuf, Altaf, and Nabeshima (2004).

help of fiscal and trade incentives and by directing publicly owned financial institutions to channel credit to provide the selected subsectors with patient capital on favorable terms.[3] The colonial authorities governing Hong Kong created the environment conducive to industrialization but allowed the private sector to take the lead (Berger and Lester 1997; Enright, Scott, and Dodwell 1997). Governments of all four economies developed and financed—with some assistance from overseas—the supporting physical infrastructure. And governments in Korea; Singapore; and Taiwan, China, actively promoted policies to deepen human capital and to build a literate and trainable workforce that could readily master new technologies and adapt to technological change. Cultural factors ensured that industrial labor in these economies also quickly acquired the discipline that is a key to industrial productivity and to moving up the ladder of industrialization.[4]

Although most other developing countries stumbled in the 1970s and several frontrunners squandered the growth momentum acquired in the preceding decade—including countries that had shown considerable promise in the 1960s, such as Brazil, Kenya, and Pakistan—the four East Asian economies proved their mettle by averaging growth rates of 9 percent per year. This performance helped forge the legend of the East Asian Tigers. During the 1980s, despite some economic turbulence, the four charter members of the society of East Asian Tiger economies consolidated their reputations and staked out an enduring niche in the annals of development.

Needless to say, the achievements of the Tiger economies did not go unnoticed, especially in their own East Asian neighborhood. Other countries were eager to develop and were casting around for models. With the benefit of hindsight, it is perhaps not surprising that neighborhood and knowledge spillover effects, certain cultural affinities, and the geography of FDI and of the emergent cross-Pacific production networks favored the Southeast Asian countries. Many other potentially well-positioned economies envied the East Asian Tigers, but only Indonesia, Malaysia, Thailand, and to a lesser extent the Philippines were able to arrive at the fortuitous combination of relative political stability and economic circumstances that served to crystallize an export-led manufacturing sector and put them on the road to lower-middle-income affluence by the 1990s. These countries emerged as the second cohort of "Tiger" economies.[5] Their natural resource–based development in the 1960s was supplemented 10 years later by industrial widening, with

---

[3] Japan's approach was described by Chalmers Johnson (1982); that of the other Tigers is approvingly discussed by Amsden (1989) and Wade (2003).

[4] These cultural predispositions have been the subject of debate ever since Max Weber (2002) singled them out and asked whether Confucian cultures had the ethical resolve needed to save and work hard in the rational pursuit of economic rewards.

[5] See Dick and others (2002); Emmerson (1999); Hill (1996); and Woo, Glassburner, and Nasution (1994) on the development history of Indonesia. See Balisacan and Hill (2003) on the Philippines and Phongpaichit and Baker (2002, 2008) on Thailand.

Malaysia in the forefront and the Philippines in fourth place. By the time these countries had advanced into the mid-1980s, manufacturing was clearly the leading sector, and exports were a major—if not always dominant—source of demand.

The East Asian crisis of 1997 to 1998 and the political changes that occurred in Indonesia and the Philippines around the turn of the century slowed growth throughout Southeast Asia. However, the persistence of this slowdown during 2000 to 2007—when gross domestic product (GDP) growth rates averaged between 4 and 6 percent compared with past rates of between 7 and 9 percent—and the steep drop in growth during 2008 and 2009 are sources of considerable concern.[6] They are sharpening fears of a decline in economic vigor and have focused attention in the Southeast Asian countries on the nature of industrialization to date, on the desirable direction of future industrial change, on the drivers of growth, and on policies to enhance the performance of current drivers as well as policies to facilitate the emergence of a new generation of industrial drivers.

The Southeast Asian Tigers feel threatened. Even though their growth rates have remained above the average for the world and also above the average for developing countries,[7] their economic performance falls short of that in the first half of the 1990s. The underlying worry is that it presages the beginning of a downward trend, the harbingers of which are lower rates of investment, persistently low rates of total factor productivity, and low levels of innovativeness. The Southeast Asian Tigers' worries motivate three questions, which this book attempts to answer. First, are the Tigers rightly threatened by a creeping economic sclerosis or what some observers are calling the *middle-income trap*? Second, if the threat is real, what are the underlying causes? Third, are there ways of neutralizing the problems and at least maintaining if not raising the growth rates of the recent past?

This book will respond to these questions by means of a comparative analysis of the Tiger economies that is centered on Malaysia. This analysis draws on a comprehensive set of techniques and indicators to assess competitive pressures, to gauge industrial and technological capabilities, and to indicate some of the directions industrial change in Southeast Asia could take. Thus, the book seeks not only to view industrial evolution in the region from a comparative perspective—taking account also of what is happening and has happened in other parts of East Asia—but also to illuminate this ongoing and uncertain process using some of the latest empirical techniques devised for this purpose.

---

[6] Malaysia, for instance, managed a GDP growth of 4.6 percent in 2008. However, the real GDP contracted by 3.6 percent between the third and the fourth quarters of 2008. GDP is forecast to contract by between 4 and 5 percent in 2009, and the Malaysian economy is not expected to fully recover from the slowdown until 2011 (Burton 2009a). To counter the downturn, the government introduced a stimulus package totaling RM 60 billion (US$16 billion, 9 percent of GDP) on March 10, 2009 (Burton 2009c; Netto 2009).

[7] During 2000 to 2007, the global economy averaged a growth rate of 3.3 percent, and developing countries as a group averaged a growth rate of 6.3 percent.

The balance of this chapter provides the developmental and international contexts with reference to which these questions will be addressed. It explains the book's preferred angles to tackling them. The chapter also outlines the contents of the volume and foreshadows the principal findings and conclusions.

Brief Development History of the Four Southeast Asian Tigers

Prior to the first stirrings of industrialization in the 1960s, the four Southeast Asian countries had not experienced preparatory protoindustrialization, lacked a class of entrepreneurs other than the ones engaged in commerce and petty trade, and had nothing resembling the deeply entrenched cultures of literacy to be found in China and Japan.[8] The wealth of these countries resided mainly in their natural resources and the development that occurred in colonial times and in the immediate aftermath of independence. It was based on exploiting minerals, timber, and the exportable bounties derived from rice and tree crops grown on fertile, well-watered land. Although timber resources are approaching exhaustion in Malaysia, the Philippines, and Thailand and the production of tin is down to a trickle in Malaysia,[9] rice, tree crop, and fisheries production remains vital to the performance and exports of these countries. Indonesia and Malaysia are large producers (and net exporters) of oil and gas, and Indonesia benefits significantly from its rich deposits of gold and copper ores. Nevertheless, as a share of GDP, of exports, and of growth, the contribution of the primary sector is well below what it used to be in all four countries, having fallen most steeply in Indonesia (see tables 1.1 and 1.2).

On the threshold of industrialization, the four leading Southeast Asian economies depended heavily on natural resources, unlike the four Tiger economies and unlike Japan in the 1950s (although Taiwan, China, had a highly productive agricultural economy, and exports of farm products remained important until the early 1990s).[10] To varying degrees, six factors shifted Southeast Asia's development trajectory and enabled the Southeast Asian economies to begin building manufacturing activities, which first supplemented and then displaced natural resource–based industries as the engines of growth and the sources of employment. Taking full account of these factors and the regional circumstances in

---

[8]Morris-Suzuki (1994) and Rawski (1979) have estimated that by the middle of the 19th century about half of the male population in China and Japan and a lesser percentage of the female population were literate.

[9]Rising prices led to a revival of tin mining during 2007 and 2008.

[10]Under Japanese occupation, the cultivation of rice, sugarcane, and vegetables in Taiwan, China, was actively promoted, and agricultural research and extension contributed to steady gains in land productivity.

Table 1.1 Export Shares of Primary and Resource-Based Products

Country	Year	Primary products (%)	Agro-based products (%)	Other resource-based products (%)
Indonesia	1970	28.4	37.7	33.4
	2006	24.8	18.4	19.7
Malaysia	1970	25.9	59.8	10.3
	2006	7.6	10.7	10.0
Philippines	1970	33.3	60.1	3.5
	2006	6.5	4.8	5.1
Thailand	1970	67.0	26.7	2.7
	2006	9.3	10.0	9.0

Source: United Nations Commodity Trade Statistics Database (UN Comtrade).

Table 1.2 Contribution of Agriculture

Country	Year	Share of agriculture in GDP (%)	Average contribution to growth of agriculture (percentage points)	Average share of agriculture growth to overall growth (%)
Indonesia	1970s	44.9	1.7	21.4
	2000s	12.9	0.4	9.5
Malaysia	1970s	26.7	1.4	3.3
	2000s	8.7	0.3	3.5
Philippines	1970s	29.5	1.1	15.3
	2000s	14.2	0.6	13.7
Thailand	1970s	25.9	1.1	13.0
	2000s	10.7	0.3	6.6

Source: World Bank World Development Indicators database.

the 1960s and early 1970s can shed light on past achievements and also illuminate the causes of current worries.

The East Asian Model

First, the economic successes notched up by the original Tiger economies in the 1970s reverberated throughout the East Asian region and exerted a profound demonstration effect. Adopting the "East Asian model" came to be viewed as the recipe for development, and each of the aspiring Tiger economies of Southeast Asia began adapting that recipe to its own political conditions and resource endowments. Second, the efforts of these economies to industrialize were aided financially and through trade policies by Japan, the United States, and other Western countries eager to support market-based economies following the culmination of the Vietnam conflict. Third, export-led growth, as demonstrated by Korea and Taiwan, China, was quickly embraced by Western donors and international financial institutions. Fourth, Southeast Asian governments came under pressure to accelerate development and thereby nip any nascent insurgencies seeking to destabilize the existing order.[11] Fifth, this objective combined with the interests of U.S. and Japanese multinational corporations (MNCs) to geographically diversify their labor-intensive assembly manufacturing to countries where wages were lower, thereby smoothing Southeast Asia's transition from natural resource–based activities to manufacturing. Finally, the process was expedited by the presence of strong, authoritarian regimes in all four countries for most of the 1970s and 1980s. Such regimes were embraced by the West and Japan because in three out of the four cases—the Philippines being the exception—they arguably quickened the transition to industrialization by acting decisively, by implementing policies (more or less effectively) to strengthen market institutions, by tying their legitimacy to achieving economic growth, and by ensuring (for the most part) law and order, which crucially underpinned rapid growth.[12]

Indonesia, Malaysia, the Philippines, and Thailand started out with similar aspirations and comparative advantages, but inevitably, given their different sizes, political systems, and cultures, each trod a different path to industrialization. Of the four, Malaysia was the least populous in the early 1970s,[13] and in per capita terms, its resource endowment, comprising cultivable land, mineral resources, and timber, was greater than that of the other countries. This feature was also reflected in its higher per capita income, life expectancy, and literacy; its share of trade to GDP; and its lower rates of poverty. Except for a period in the latter half of the 1960s, when ethnic tensions flared, Malaysia was also the most politically stable of the four countries, and this stability extended to the macroeconomic sphere.

After Suharto took over the reins of government in 1967, Indonesia settled into a long period of social and political stability, and after the early 1980s, the country enjoyed almost two decades of growth and price stability. Frequent changes of government in Thailand[14] and, to a lesser extent, in the Philippines had only a limited effect on the social stability and economic performance of the former but lessened the effectiveness of policy making in the latter.[15] The Philippines

---

[11]Smallholder-based oil palm growing schemes managed by a newly created Malaysian agency, the Federal Land Development Agency, were also aimed at lessening the risk of unrest. The schemes gave peasants a larger stake in the economy and a means of improving their lots, because returns from oil palm cultivation were high.

[12]On Indonesia, see Hill (1996) and Woo, Glassburner, and Nasution (1994). On Thailand, see Phongpaichit and Baker (2002).

[13]When it gained independence in 1957, Malaysia's population was just 7.4 million (Yusof and Bhattasali 2008).

[14]Since 1932 (the year when the absolute monarchy was abolished), there have been 18 coups in Thailand, of which 11 have been successful (Pilling 2009).

[15]The Philippines experienced a number of boom-bust cycles since the 1960s, unable to sustain growth of 5 percent or more for a long period of time (Bernardo and Tang 2008).

experienced fairly severe macroeconomic stresses in the first half of the 1980s as a consequence of the oil shock and lapses in policy—problems that the other three countries were able to contain. When President Fidel V. Ramos took the office in 1992, he introduced a number of reforms, mostly in line with the Washington Consensus, to put the Philippines on a more stable growth path (Bernardo and Tang 2008). What is notable over the entire period from the early 1970s to the present is the degree to which all four countries adopted variants of a common growth strategy, which was to develop export-oriented manufacturing and agricultural activities with the help of FDI and to support these activities with trade and other incentive regimes, infrastructure building, and investment in education. Manufacturing, agriculture, and construction emerged as the sectoral drivers of growth, with services playing a bigger role in the Philippines relative to manufacturing (Balisacan and Hill 2003).[16] The strategies followed had enough in common for this analysis to use the experience of Malaysia as a mirror for the experience and prospects of the three other countries.

The Malaysian Experience

To jumpstart the growth of the manufacturing sector, Malaysia, much like Singapore, needed firms that could compete in the export market and quickly ramp up their sales to the industrial countries. Because few, if any, domestic firms had such potential, Malaysia had to provide the incentives to attract foreign manufacturers; it did so effectively, starting with the establishment of a free trade zone in Penang in 1971. MNCs, initially from Japan and the United States and later from Europe, were offered attractive terms, were satisfied by the quality of the labor force for low-tech assembly operations, and were provided the physical infrastructure to meet their requirements. The government's control over political freedoms and firm grip on society[17] maintained the peace, and the business climate was equal to if not better than that of other Southeast Asian countries.

---

[16]The Philippines was able to attract some back-office process outsourcing, accounting for 15 percent of the global market share in 2008 (Gott 2009). Although this outsourcing has generated employment, the links to other industries are limited and did not stimulate any production activities (Magtibay-Ramos, Estrada, and Felipe 2008). The Philippines has also developed a sizable animation industry (total revenues of US$110 million in 2008), which contracts with multimedia, moviemaking, and animation firms in Japan and the United States. Approximately 10,000 workers are directly employed in the animation industry. Although this industry has faced difficulties since 2001, the government and the newly created industrial association are collaborating to facilitate its growth by establishing information technology zones, hosting animation festivals, encouraging international marketing, and setting up training courses.

[17]The Internal Security Act, which was introduced following serious rioting in 1969, gave the government broad powers of arrest and detention.

Industrialization and the Changing Export Mix

The Malaysian state's commitment to industrialize and to invest in the supporting infrastructure, the buoyant performance of the primary sector, and the inflow of FDI, boosted Malaysia toward a higher growth path. By the end of the 1970s, Malaysia was a substantial exporter of raw materials, textiles, other light manufactures, and electronic and electrical products assembled in factories established by MNCs in Penang and in the Kuala Lumpur–Port Klang area. Manufactures first complemented and then overshadowed resource-based exports. As the growth rate of GDP soared, the stage was set for a virtuous spiral, with rising incomes pulling up savings and helping to finance high rates of investment, both public and private. FDI in manufacturing generated some vertical links, which led to the widening of industrialization, and the Malaysian government's own efforts (like those of the Indonesian government) at targeted development of the transport, engineering, metallurgical, and petrochemical industries further diversified the base of manufacturing. The dramatic change in the industrial composition of the Malaysian economy can be seen from the changing export shares presented in table 1.3. In 1970, more than 95 percent of Malaysia's exports were resource based, as were the exports of the three other Southeast Asian comparators.[18] Ten years later, the share had declined to 83 percent, and it had also fallen in the Philippines and Thailand. By 1990, only about one-half of Malaysia's exports were natural resource–based products, a smaller share than those of its comparators. By 2007, the share of such exports was less than 30 percent in Malaysia as well as in the Philippines and in Thailand (2006). Only Indonesia remains a large exporter of primary products and other resource-based commodities, with these accounting for over 60 percent of its exports (see table 1.3).

Innovation Capabilities and Dynamism of Firms

During the period extending from the mid-1970s through the mid-1990s, Indonesia, Malaysia, and Thailand achieved their growth objectives assisted by FDI[19] and a sustained increase in manufactured exports. Export growth slowed in the mid-1990s because of rising costs and competition from China and because Southeast Asia was slow to diversify into higher-value products. The crisis in 1997 to 1998 was partly caused by weakening trade performance and dampened investment and growth. Although growth revived within a few years, it did not climb back to earlier trend rates. Nevertheless, the long boom experienced by

---

[18] Yet earlier, rubber accounted for 70 percent of Malaysia's exports in 1948 to 1952 and for two-thirds of exports during 1963 to 1967 (Lim 1973).

[19] Following the Plaza Accord in 1985, which resulted in the appreciation of the Japanese yen relative to the U.S. dollar, the flow of Japanese FDI to Southeast Asia increased.

Table 1.3 Export Structure of Southeast Asian Countries, 1970–2007

Country	Year	Electronic and electrical	Other high technology	Textile, garment, and footwear	Other low technology	Automotive	Process	Engineering	Primary products	Agro-based products	Other resource-based products
Indonesia	1970	0.47	0.30	0.25	0.07	0.02	0.03	0.03	80.53	6.07	12.75
	1980	0.65	0.07	0.61	0.18	0.16	0.40	0.76	85.31	10.51	2.41
	1990	12.90	0.28	12.28	3.64	0.81	4.62	4.52	56.04	17.53	4.03
	2000	6.11	0.45	15.03	8.25	1.88	5.47	5.81	31.99	14.62	5.95
	2007		0.50	9.58	6.29		4.51		36.33	17.97	11.02
Malaysia	1970	0.27	0.41	0.78	0.85	0.48	0.52	0.82	64.61	27.56	3.70
	1980	8.89	0.94	2.30	1.18	0.08	1.36	2.29	53.16	28.26	1.55
	1990	25.70	1.85	5.95	5.46	0.42	2.13	9.96	26.05	21.11	1.38
	2000	53.86	1.72	3.45	6.21	0.43	3.31	10.19	10.84	8.76	1.21
	2007	40.65	2.63	2.75	8.66	0.66	5.15	9.99	15.90	11.80	1.82
Philippines	1970	0.00	0.01	0.75	1.42	0.00	0.54	0.14	14.32	60.76	22.05
	1980	1.31	0.35	9.01	5.96	0.68	1.46	1.16	17.89	39.10	23.08
	1990	13.12	0.22	16.30	11.41	0.52	5.03	4.74	20.59	20.12	7.95
	2000	67.02	0.80	8.46	3.45	1.69	0.71	9.23	3.58	3.77	1.29
	2007	48.16	0.99	5.80	4.86	5.66	1.78	8.52	10.23	9.33	4.67
Thailand	1970	0.04	0.13	1.48	0.66	0.25	0.53	0.47	80.59	12.63	3.22
	1980	0.33	0.38	8.95	2.75	0.24	2.90	5.97	63.25	9.54	5.68
	1990	16.15	0.37	20.45	10.55	0.57	3.36	6.44	24.03	12.79	5.29
	2000	30.45	1.31	10.49	11.24	3.81	3.58	12.29	12.68	9.81	4.34
	2006	25.20	1.57	12.17	9.16	6.62	6.58	15.49	10.35	8.42	4.43

Source: UN Comtrade.

these economies transformed their economies structurally, and they have all acquired considerable manufacturing capabilities, most notably in electronic and electrical engineering, textiles, and the automotive industry. No more than a handful of other developing economies have achieved as much in 35 years. But unlike the original East Asian Tiger economies, the Southeast Asian Tigers have yet to build the indigenous capacity to design, to innovate, and to diversify into new and more profitable areas with good long-run prospects, and very few of their firms have created regional—much less global—brand names. They are adept at assimilating technologies from overseas and have the production and plant management skills to match the labor productivity levels of industrial countries in the production of standardized commodities. However, innovation—product or process—remains mainly a preserve of the MNCs; indigenous firms do very little innovation. More disquieting is the sparseness of backward links from MNC operations, which would signify progressive industrial deepening, as has occurred in Korea and Taiwan, China, and as is already under way in China. This lack of backward links means that domestic value added in manufacturing remains low. Moreover, none of these countries has nurtured large and dynamic producers of tradable services, à la India in information technology and à la Hong Kong, China, and à la Singapore in finance, logistics, and other business services.

In Malaysia, efforts to promote domestic industrial links began in the early 1990s, and for more than a decade, the government has attempted to improve the quality of skills and to encourage research and development (R&D) so as to induce innovation and to stimulate value-adding product differentiation by Malaysian firms. The three other countries have taken smaller steps in these directions. Because tangible evidence of an emerging culture of innovation has been slow to surface, governments—aware of rising competitive pressures—worry that the virtuous spiral of the recent decades might dissipate and view declining rates of investment as a sign that the business communities are finding that profitable opportunities are becoming scarce.

A closer look at the Malaysian case and a contrasting of Malaysia's development with that of Korea and Taiwan, China, suggests four reasons Malaysian industry, much like the industries of other Southeast Asian countries, has proven thus far to be less dynamic and innovative. First is the long-standing reliance on MNCs to provide Malaysia with export-oriented platforms and the inability to induce foreign companies to forge and deepen backward links and to generate knowledge spillovers by seeking out and closely involving local suppliers in the manufacture of increasingly more complex components and subassemblies. Although the industrial presence of MNCs is significant, their knowledge footprint has remained small. They have made Malaysia and the other Southeast Asian countries wealthier, but their contribution to development is far less. The Malaysian authorities had an opportunity to multiply links arising from the presence of MNCs, but it was allowed to slip, in part because rules requiring local sourcing were not enforced for fear of driving the MNCs away. But other reasons—associated with the quality of skills, for example—explain Malaysia's inability to exploit the presence of MNCs.

Second, Malaysia (unlike Korea and Taiwan, China) has been slow to try to construct a culture of excellence and innovation through its schools and tertiary-level institutions. Korea; Singapore; and Taiwan, China, saw the need for such a culture and took steps to make it a reality, recognizing that doing so might take decades. They emphasized the quality of STEM (science, technology, engineering, and math) and other skills, and through a combination of foreign schooling and steady improvement in the training provided by local universities and vocational training facilities, they raised the caliber of the workforce. The governments in these economies, with the backing of the business communities, made the pursuit of innovation through research central to their strategies of sustaining industrial competitiveness. Both governments have urged industry to ceaselessly evolve so as to retain its competitive edge and have supported exhortations with tangible incentives. This awareness of external competition has been internalized by the business sector, and to a greater degree than in other industrializing countries, industry in Korea and Taiwan, China, has worked closely with the government to retain a competitive edge. In a globalized world, a duck that remains stationary is a sitting duck for competitors. The rules associated with the New Economic Policy (NEP) constrain Malaysia's efforts to raise the quality of its student body, teaching, and research, but the government has been less than resolute and less than creative in easing the rules that bind the administrators of public universities and prevent universities from competing, from recruiting talented faculty, and from taking the difficult steps to enhance student performance.[20] However, the other three countries, although less subject to social pressures, have had no greater success in raising the quality of tertiary education and in promoting research.

Third, in Malaysia and the three other Southeast Asian countries, the volume of entrepreneurship and its creativity are not sufficient to raise the tempo of innovation. The quality of education and skills is one factor, as previously noted, but industrial organization, the structure of ownership, and the competitiveness of the market environment are also of relevance. Major conglomerates—either family owned or controlled by well-connected business elites—dominate the industrial landscape, and the high level of concentration raises entry barriers for new firms and inhibits the growth of existing firms that could challenge the incumbents

---

[20]See World Bank (2007c) for an assessment of tertiary education in Malaysia and for views on what it takes to build world-class universities.

(Unger 1998).[21] The extremely low turnover of the leading firms in Malaysia and in the other Southeast Asian countries and the few additions to the ranks of the largest business entities reflect unfavorably on the dynamism of the business environment.

Fourth, the importance of government-controlled corporations (GCCs) in Indonesia, Malaysia, and to a lesser extent the two other countries further erodes dynamism and innovativeness. The incentives impinging on GCCs do not predispose them to strive after competitiveness by way of innovation. And it is rare to find a public corporation (anywhere in the world) whose organization, management system, and incentives place a high premium on initiative, innovativeness, and risk taking. The culture of GCCs the world over discourages risk taking and rarely rewards innovation, which can disrupt day-to-day operations, reduce the sales of existing products, and require additional effort at marketing and customization. In fact, few public entities are driven by their principal shareholder to maximize profit—satisfactory profits are usually enough.[22]

In Malaysia, the dynamism of the market environment is further affected by the oligopolistic structure of business. Even though formal barriers to trade, for example, are low and the business climate is considered to be moderately good, the dominance of a few large, privately owned firms and of government-controlled enterprises—all closely linked with powerful government agencies—dampens competition from new entrants or foreign firms. Consolidation of Malaysia's domestic banking system under government guidance after the crisis of 1997 and 1998 has reduced the number of core banks to just six, of which two are government owned (Gomez 2006).[23] Given the relatively underdeveloped financial market—especially the lack of alternatives to loans—this consolidation also limits the entry of firms.

The introduction of the New Economic Policy in 1991 led to the emergence of a Malay business elite owning large enterprises. Many Malaysian conglomerates used a holding company structure to own firms and to grow the business group through cross-shareholdings and interlocking directorships, where two or more

---

[21] Family ownership and the dominance of family-controlled conglomerates are also found in the three other Southeast Asian countries. Family ownership can have advantages because it provides firms with decision-making flexibility and increases the speed of response to opportunities, but as Studwell (2007) has pointed out, few of the heads of these conglomerates—or even the generation now succeeding them—are showing much industrial entrepreneurship.

[22] There is a vast literature on the woes of the public sector. Much of it argues for privatization, especially of manufacturing firms. For a survey, see Yusuf, Nabeshima, and Perkins (2005).

[23] Policies announced in April 2009 promise some loosening of the entry barriers to majority foreign-owned investment in banks ("Malaysia: Najib Eases" 2009).

firms shared members of the board of directors (Gomez 2006).[24,25] The NEP also introduced the Industrial Coordination Act in 1975, requiring firms to obtain licenses once they had 25 or more employees (or paid-up capital exceeding RM 250,000); this legislation was used to control entry of firms and to ensure that firms complied with the objectives of the NEP (Lee 2007). However, despite support for the development of Bumiputra firms, which continued under the New Development Policy (the policy that superseded the NEP in 1991), among the top 10 firms in 2001 in terms of domestic market capitalization, 7 were government-owned firms, while the rest were Chinese owned. In addition, none of the top 50 Malay firms listed on the Kuala Lumpur Stock Exchange was successful in expanding its business in manufacturing. Instead, the focus of these firms is on finance, construction, property development, and telecommunications, where government patronage is much stronger (Gomez 2006).

Throughout the world, the innovation driving competitiveness and profitability largely depends on the strategies and decisions of firms. Governments, research entities, and universities can facilitate and provide the institutions, the talent, and some of the resources, but firms are the ones that do the applied research and

---

[24]Business groups exist in many different countries. Some of the notable ones are the *business houses* in India, *keiretsu* in Japan, *chaebols* in Korea, and *grupos económicos* in Latin American countries (Chang 2006a). In East Asia, business groups in one form or another are pervasive, and they account for a large share of economic activities. In Korea, the top 30 chaebols accounted for 40 percent of industrial output in 1996. Thirteen of the weaker ones subsequently went bankrupt after the Asian crisis of 1997 to 1998 (Chang 2006b). Large business groups accounted for 24.3, 24.9, and 34.6 percent of the market capitalization in Thailand, Malaysia, and Singapore, respectively, in 2002. Many of these business groups were supported by governments pursuing their economic and industrial development objectives. Before the crisis, the top five business groups in Thailand were all connected to main banks. In addition, those firms were mainly controlled by a founding family or families. After the crisis of 1997 to 1998, the finances of a quarter of the families were badly damaged, and some of these families lost their controlling shares in banks (Phongpaichit and Baker 2008; Polsiri and Wiwattanakantang 2006). However, even after the crisis, firms that are politically connected (and globally oriented) still outperform those that are not (Imai 2006). Similarly, many of the business groups in Indonesia owned banks as a means of mobilizing funds and gaining access to credit (Hanani 2006). However, the relationships with governments were more arm's length in Japan, Korea, and Singapore than in Indonesia, Malaysia, and Thailand—although in Korea chaebols have actively blocked attempts by governments to enact competition policies (Kang 2005). In these latter countries, the close connections with governments were necessary for the success of businesses and, in the event of a crisis, for their survival (Chang 2006a).

[25]Accounting reform in Japan—which forced firms to disclose the market value of equity holdings—compelled these firms to sell off their stockholding in other firms (cross-shareholding) and in banks (Ahmadjian 2006).

through trial and error actually determine whether innovation succeeds. One of the most durable findings of economic research is that markets dominated by oligopolies—and, in particular, by public corporations—will be less open to change and innovation.

The Southeast Asian Tiger economies have certainly acknowledged the value of openness and competition. In fact, their performance is in no small part the outcome of their efforts to improve the business climate. Table 2.A.5 in the annex to chapter 2 presents the rankings of the four countries in 2007 and again in 2008. Openness and competition have tended to be uneven: the tradable sectors dominated by MNCs are undoubtedly open and competitive, but in other parts of these economies, competitive pressures are weak because of the salience of public corporations or private monopolies or because the goods or services are not tradable. It is difficult to delineate the gradations of competition, but casual empiricism would suggest that they are significant.[26] They provide the larger companies—which tend to be diversified—with a cushion, and by ensuring a relatively secure stream of revenue, they diminish the need to pursue more entrepreneurial strategies that would emphasize innovation.

Analyzing Industrial Change in Southeast Asia

In broad strokes, this discussion defines the strategies, the achievements, and the weaknesses of the Southeast Asian Tigers and indicates why they feel vulnerable in a world where the future trend in global trade is uncertain, where new competitors have emerged, and where FDI might be less abundant and might not provide a stepping-stone to the next stage of development. Under these circumstances, the Southeast Asian Tigers are having to reassess their acquired comparative advantages in traditional resource-based products and in newer manufactures and services and are having to search for strategies that will prove equal to the new challenges.

Recent research provides a number of techniques for evaluating a country's economic capabilities and competitiveness relative to its trading partners. Guided by the availability of the needed data, this study marshals a variety of these measures to assess manufacturing and technological capabilities in Malaysia. Where possible, it compares these capabilities with those of other Southeast and Northeast Asian economies. These measures show how comparative advantage is evolving and the likely areas of competitiveness in the future. By highlighting the strengths—and also the vulnerabilities—of the Malaysian economy in relation to other East Asian economies, the analysis defines Malaysia's options and suggests

---

[26]Cassey Lee (2007) reviews the state of competition and competition policy in Southeast Asian countries, but hard empirical details on concentration and other attributes of competition are virtually nonexistent.

how current policies might be modified and supplemented so as to achieve rates of growth that are acceptable and also sustainable. This approach also illuminates the challenges facing the other Southeast Asian countries and points to policies of relevance to them as well.

The measures employed use trade data to analyze industrial change and competitiveness; use FDI and domestic investment to determine what the market perceives to be the comparative advantages of the Malaysian economy; use data on R&D and on patenting to gauge innovativeness and the focus of technological advance; use information on the supply of skills and the numbers of research universities and institutes to show the scale of the national innovation system; indicate how much benefit growth and innovation could obtain from exploiting agglomeration economies inherent in the major urban centers; use the growth in total factor productivity to capture at the macrolevel how much improvement Malaysia has derived from efficiency and technological gains over the past two decades; tap the indexes computed by the World Economic Forum, the World Bank, and other bodies to track the perceived changes in competitiveness and in innovation readiness; and then, with the help of simple models, forecast Malaysia's economic prospects on the basis of the performance of its trading partners, trend rates of past GDP growth, and movements in key variables. Taken together, these measures comprehensively assess the capabilities and potential of the Malaysian economy and provide a sense for where future comparative advantage may lie in a changing global environment.

The findings and indicators assembled in this book lead to recommendations for policies affecting the direction of industrial development, FDI, R&D spending and its composition, the geography of industrialization, and the building of human capital. Better policies in each of these areas could reinforce industrial competitiveness, stimulate innovation, and ease worries over the pace of future development. However, to derive the most mileage from sector-specific policies, Malaysia and the other Southeast Asian Tigers will also need to pursue, in parallel, sociopolitical policies and policies affecting competition and industrial organization. For example, raising the quality of human capital and creating an innovation culture require more than just increased spending on education and research. The society must offer opportunities for all; handsomely reward talent, risk taking, and innovation; and encourage bright minds to inquire, analyze, debate, and through this process arrive at solutions to current problems and to problems on the horizon.

Tiger in the Spotlight

Why single out Malaysia from among the four Southeast Asian countries for special attention? The reasons are straightforward. Malaysia leads the field with respect to per capita GDP and manufacturing capabilities in key industries such as electronics, and with reference to technological expertise in palm oil and rubber.

Its ethnic heterogeneity is potentially advantageous from the standpoint of entrepreneurship and idea generation. Low levels of poverty are another asset.[27] Being the smallest in terms of population, Malaysia is also the most dependent on the international market. Among the Southeast Asian countries, its manufacturing activities are at the stage in their lifecycles at which the transition to higher value-adding activities is potentially feasible apart from being necessary to maintain the tempo of growth. Because it is in the lead, the Malaysian Tiger is the most exposed to threats from competitors at equivalent and higher levels of the food chain. Malaysia is also arguably the one Tiger that can show the others a way of climbing up the food chain. Malaysia feels most urgently the pressure of competition from exporters elsewhere in Asia. But it also has the base of accumulated learning and infrastructure and can mobilize the financial resources, which are among the necessary ingredients for the next stage of development.

This book is organized as follows. Chapter 2 provides a brief overview of the evolution of the Malaysian economy. The key to understanding the growth prospects of an economy is to gauge comparative advantage. Chapters 3 and 4 explore the comparative advantage of Malaysia by using extensive sets of tools based on disaggregated export and import data. This examination is followed by a detailed analysis of three key industrial subsectors: (a) electronic components, (b) automotive parts, and (c) palm oil and food products (chapter 5). Chapter 6 examines the innovation capability of Malaysia, also by using a variety of measures. Chapter 7 assesses the long-run growth trends in Malaysia, and chapter 8 asks whether Malaysia and other Southeast Asian economies can grow at a similar pace as in the past. The final chapter (chapter 9) offers a menu of policies to help Malaysia move toward the next stage of industrial development.

---

[27] By the most recent count, poverty, measured at US$2 per day, affected about 0.5 percent of the population, down from 2.0 percent in 1995.

2

Malaysia
The Quintessential Maturing Tiger Economy

Malaysia gained independence in 1957. At that time, agriculture and mining accounted for more than 45 percent of gross domestic product (GDP) and 66 percent of employment. More than three-quarters of Malaysia's export earnings derived from rubber and tin (Khalafalla and Webb 2001; Zakariah and Ahmad 1999). A decade later, Malaysia embarked on a strategy to enlarge the share of manufacturing in GDP and to substantially augment the contribution of manufactured exports to growth. The new emphasis on industrialization was formalized by the Investment Incentives Act of 1968 and the establishment of the Malaysian Industrial Development Authority (MIDA). In 1971, this legislative step was reinforced by the introduction of the New Economic Policy (NEP), which underscored the role of manufacturing (Zakariah and Ahmad 1999), and by the Free Trade Zone Act of 1972, which signaled the start of an effort to change the composition of exports (Rasiah 2008; Yusof and Bhattasali 2008). This effort was reinforced by the Promotion of Incentives Act of 1986.

Sources of Growth

Throughout the early 1980s, steadily rising exports propelled the expansion of agriculture and manufacturing; however, domestic demand remained the principal source of economic growth. From about 1983 through the mid-1990s, Malaysia's growth was mainly export led, with manufactured products coming to dominate the export mix.[1] By 1995, a third of GDP was sourced from the

---

[1] The beginnings of Malaysia's electronics industry can be traced to a factory established by Matsushita Electric in Shah Alam in 1965. Later, the electronics industry took root in Penang when locally owned Penang Electronics commenced operations in 1970 (Rasiah 2007, 2008).

Table 2.1 Historical Performance of Malaysia

Indicator	1960s	1970s	1980s	1990s	2000–06
Average GDP growth (%)	6.6	7.7	5.9	7.3	5.2
Average per capita GDP growth (%)	3.5	5.2	3.2	4.5	3.2
Average export growth (%)	6.0	8.2	9.2	12.7	7.2

Source: World Bank World Development Indicators database.

Table 2.2 Average Sectoral Contribution to Growth: Demand Side

Period	Contribution (%)			
	Consumption	Government spending	Investment	Net exports
1991–94	39.2	11.7	72.3	−23.2
1995–2000	38.3	7.5	8.4	45.8
2001–06	66.7	27.7	7.4	−2.1

Source: Bank Negara Malaysia.

manufacturing sector, which employed one-quarter of the labor force. Both agriculture and mining lost ground: their combined share of GDP and employment declined to 21 percent and 18.5 percent, respectively (Economist Intelligence Unit 1996). In the space of 25 years, the structure of Malaysia's economy had been transformed, with manufacturing and urban services displacing natural resource–based products. Largely because of this transformation, GDP growth averaged more than 6 percent per year during the 1960s to 1990s, and per capita GDP rose to US$3,510 by the mid-1990s (see table 2.1).

Since 2000, Malaysia's average rate of GDP growth has been more than 2 percentage points lower than it was during the 1990s (see table 2.1). In addition, between 1995 and 2000, the contribution of the increase in net exports to growth oscillated but was still significant (see also Haltmaier and others 2007), whereas between 2001 and 2006, the contribution of net exports to growth was negative (see table 2.2).[2] A steep decline in private investment further dampened economic performance.[3] Partially offsetting these two flagging drivers of growth, domestic

---

[2] Khalafalla and Webb's (2001) tests using Granger causality support earlier findings of export-led growth during 1965 to 1980. But in 1981 to 1996, growth and imports "cause" exports. The data in table 2.2 also show that net exports were not a source of growth from 1991 to 1994.

[3] DeLong and Summers (1991, 1993) showed that investment in equipment was causally related to growth through the increase in production capacity, productivity, and other spillover effects. Mazumdar's (2001) research supported this finding by drawing attention to the effect of imported equipment on growth in developing countries. Her results were reinforced by Mody and Yilmaz (2002), who determined that imported equipment contributed to export competitiveness and growth in industrializing countries by reducing costs and spurring domestic innovation.

Table 2.3 Average Sectoral Contribution to Growth: Supply Side

Period	Contribution (%)			
	Primary	Manufacturing	Construction	Services
1991–94	3.2	31.6	5.3	60.0
1995–2000	9.0	45.2	1.8	44.0
2001–06	8.8	38.0	0.5	52.7

Source: Bank Negara Malaysia.

consumption and public investment have risen robustly. Much of the growth between 2000 and 2006 can be traced to those two sources of demand, and this trend continued in 2007 ("Malaysia: Domestic Demand" 2007). The Malaysian economy had been pulled by four locomotives. Now two of them—net exports and private investment—are far weaker.

A sectoral decomposition of growth in table 2.3 indicates that in Malaysia the service sector, rather than industry, is pulling the economy, as it does in most middle- and high-income countries. The implications for productivity are mixed. In principle, because the gap in labor productivity between Asia and the United States is higher in services, the scope for growth—through catch-up and more effective harnessing of information technology (IT) through organizational changes and training—is also greater, especially in areas such as retailing and finance (Jaumotte and Spatafora 2007). However, the record of long-term productivity growth strongly favors the manufacturing sector as the main contributor to productivity growth. From 1995 to 2000, manufacturing's share of Malaysian GDP slightly exceeded that of services, but as the share of the manufacturing sector diminishes, so will its contribution to total factor productivity (TFP) growth.

Pinning down the causes of the decline in private investment in Malaysia and other East Asian economies between 1997–98 and 2007–08 has not been easy. In part, it is viewed as an adjustment to overinvestment in the 1990s, mainly in real estate and infrastructure supported by expansionary monetary policies and ambitious government-sponsored projects. Partly it is related to declining profitability of certain lines of business as export profit margins have been squeezed by softening prices—the result of intensifying competition—and by rising costs of labor and more recently of raw materials.[4] In addition, the weakening of investment reflects Malaysia's limited success in diversifying its industrial activities through technological assimilation into products or services

---

[4]Guimaraes and Unteroberdoerster (2006) show that in the four years prior to 1998, Malaysia may have been overinvesting to the tune of 10 percent of GDP. They also find that corporate profitability and market valuation indicators both fell from 1998 to 2004, which discouraged private investment. A similar decline in corporate returns explains the lower invstment rates in the Republic of Korea since 1997 to 1998 (Kinkyo 2007).

Table 2.4 Contributions of Input Growth and TFP Growth

Period	Gross output growth	Relative contribution to input growth (%)			TFP growth (percentage points)
		Capital	Labor	Intermediate inputs	
1971–79	8.3	32	41	27	0.9
1980–89	7.6	36	34	30	1.4
1990–99	9.9	33	28	39	0.9
2000–02	5.7	33	24	43	0.6
1971–02	7.1	34	29	37	1.0

Source: Mahadevan 2007b.

promising higher returns. Instead, the increasing specialization of East Asian producers in overlapping product categories and the enormous expansion of production capacity in China may be affecting the perception of medium-term opportunities. The estimates of growth constructed with reference to factor inputs and gains in TFP strengthen this impression (see table 2.4). As is typical for developing economies, more than 30 percent of Malaysia's growth was the result of increments in capital and intermediate inputs during 1971 to 2002, 25 percent from increased labor inputs, and 14 percent from gains in productivity, with TFP growth accounting for no more than 1.5 percentage points per year (Mahadevan 2007b).[5] Other estimates assign a higher share to TFP from 1991 to 1995—2.5 percentage points per year, close to one-third of GDP growth (see table 2.5). Malaysia's TFP growth had diminished to 1.1 percent per year in 1996 to 2000 and rose only fractionally to 1.3 percent from 2001 to 2005, considerably lower than the rate achieved by China and lower than TFP growth in Thailand (see table 2.5). In other words, capital remains the primary determinant of growth. Furthermore, because resource transfers from the primary sector to industry and services have slowed (now that only 13 percent of the workforce is employed in agriculture), and because investment spending has slackened, the gains in TFP arising from a reallocation of labor from low- to high-productivity sectors have diminished (Jaumotte and Spatafora 2007).[6] To a lesser degree, so also have the

---

[5]The experience of Thailand also shows that 90 percent of growth between 1980 and 2002 was from factor accumulation, mainly capital investment. Investment accounted for 70 percent of the growth during this period (Warr 2007).

[6]A breakdown of the sources of TFP growth in Thailand shows that the reallocation of resources from agriculture to manufacturing was responsible for 24 percent of the increase in TFP. Increasing agricultural TFP also contributed 5 percent of the growth, but the overall TFP growth was responsible for only 10 percent of growth because of negative TFP growth in industry (−7 percent) and services (−12 percent) (Warr 2007).

Table 2.5 TFP Growth for Malaysia and Other Countries

Country	Period	TFP (%)
Malaysia	6th plan period (1991–95)	2.5
	7th plan period (1996–2000)	1.1
	8th plan period (2001–05)	1.3
	9th plan period target (2006–10)	2.2
	1960–80	1.2
	1980–2000	0.5
Thailand	1977–2004	1.0
	1997–96	1.6
	1999–2004	2.1
China	1978–2004	3.8
	1978–93	3.6
	1993–2004	4.0
India	1978–2004	1.6
	1978–93	1.1
	1993–2004	2.3

Sources: Data for Malaysia are from EPU (2006) and World Bank (2005b). Data for Thailand are from Bosworth (2005). Data for China and India are from Bosworth and Collins (2007).

benefits from technology embodied in the latest equipment and the "learning" associated with increased production.[7]

Malaysia's economy is at a critical juncture, and the worldwide downturn in economic activity that gathered momentum after the third quarter of 2008 has sharpened the need for a longer-term strategy to revive and sustain growth. As noted in chapter 1, Malaysia is in the front ranks of the industrializing and rapidly growing middle-income countries,[8] but the future direction of industrial change

---

[7] These transfers of resources among sectors have boosted productivity in East Asia and will remain important for countries with a large agricultural workforce (see Jaumotte and Spatafora 2007). The experience of the United States between 1960 and 2005 can serve as a guide to the potential sources of productivity growth for Malaysia during the catch-up phase of development. The findings of Jorgenson and others (2007) indicate that five activities with the highest contribution to the growth of TFP in the United States during 1960 to 2005 were, in the order of importance, (a) production of computers and office equipment, (b) wholesale trade, (c) real estate, (d) electronic components, and (e) retail trade. At the same time, productivity growth in the United States was dragged down by negative gains in construction, oil and gas exploration, business services, private hospitals, and insurance. However, in Malaysia, TFP growth was fastest in the chemicals and vehicles industries, and TFP growth in the electronics industry has been mediocre (Kim and Shafi'i 2009).

[8] Joseph Stiglitz (2007) observes that in the 50 years since independence, Malaysia's per capita GDP adjusted for inflation increased more than sevenfold, even though the population tripled.

for these economies is increasingly uncertain. This uncertainty is underscored by a number of conditions that are specific to Malaysia but that in varying degrees are also being encountered by other Southeast Asian countries:

- Global competition in key tradable items is intensifying.
- Export value added is stagnating, while export diversification (by product or destinations) is slowing.
- The technological ranking of exports has remained largely unchanged for well over a decade.
- Foreign direct investment (FDI) is neither widening nor deepening the export product mix. It is also not serving to increase domestic value added through backward and forward links. And FDI is not helping to significantly enhance research capacity. Moreover, FDI fell in 2008 and again in 2009, and a shrinking of multinational corporation (MNC) operations in Malaysia is in the cards.
- Domestic innovation capability is increasing little if at all despite higher spending on research and development (R&D) and some widening of applied research by MNCs.

Given these conditions, will the manufacturing subsector set the pace of Malaysia's growth? If so, how would the composition and technological capability of manufacturing need to evolve? Will future economic growth depend on an accelerating transition to an economic structure in which the share of services (tradable and nontradable) and service exports is considerably larger than it currently is? To address these questions, one must take stock of where the Malaysian economy is, of recent trends affecting its development, and where Malaysia ranks in competitiveness and productivity. Malaysia is at a stage of development in which the contribution of TFP to growth performance should be increasing.[9]

Evolution of the Manufacturing Industry in Malaysia

In 1981, industrial production in Malaysia was dominated by low-tech and resource-based products such as food, textiles, wood products, petrochemicals, and rubber (see figure 2.1). Electric machinery (which includes electronics and IT-related products) accounted for close to 15 percent. With the inflow of FDI, the share of electric machinery expanded to more than 20 percent of industrial output in 1990. Gradually, the industrial output has shifted from low-tech and resource-based goods to medium- and high-tech goods. By 2002, electric and general machinery accounted for close to 40 percent of industrial output. The importance of food, rubber, and wood has diminished significantly since 1981.

---

[9]Hulten and Isaksson (2007) ascribe much of the development gap between low- and high-income countries to differences in productivity.

Figure 2.1 Industrial Composition by Type of Manufacture, Malaysia, 1981, 1990, and 2002

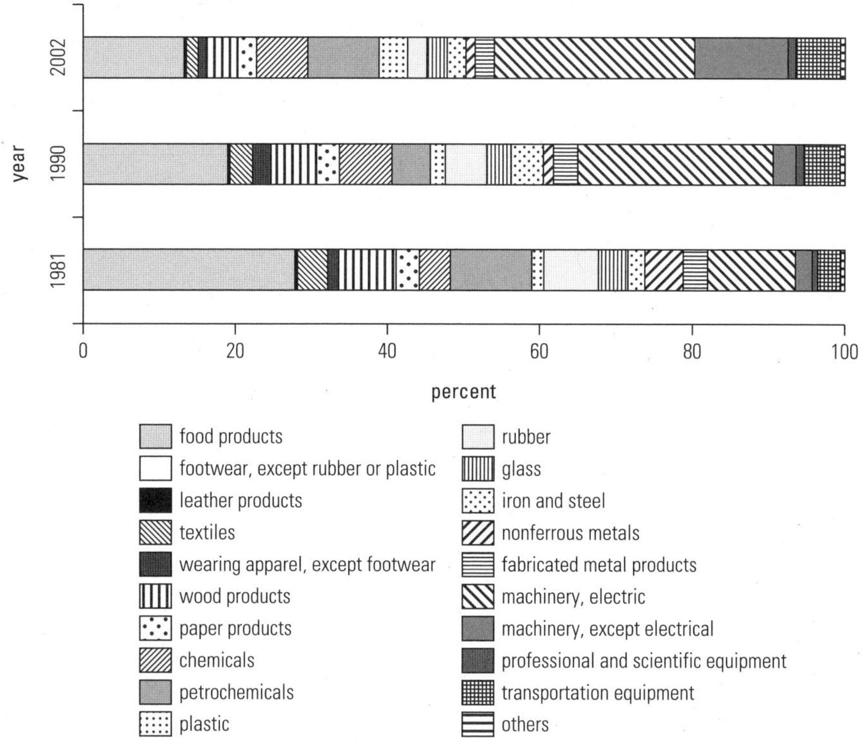
year
2002
1990
1981
0
20
40
60
80
100
percent
food products
footwear, except rubber or plastic
leather products
textiles
wearing apparel, except footwear
wood products
paper products
chemicals
petrochemicals
plastic
rubber
glass
iron and steel
nonferrous metals
fabricated metal products
machinery, electric
machinery, except electrical
professional and scientific equipment
transportation equipment
others

Source: United Nations Industrial Development Organization Industrial Statistics Database (UNIDO INDSTAT3).

The transportation subsector saw respectable growth during this period but was overshadowed by growth in electric and general machinery.

The increase in Malaysia's domestic production is reflected in the increase in its world share for these products. In 1981, Malaysia was an insignificant player in electronics and machinery, but by 2002, it emerged as a main player (although smaller than China) in these products (see table 2.6).[10]

Most industrial subsectors experienced a healthy labor productivity growth (measured by changes in value added per worker). Average annual growth was highest for the chemical and petrochemical industries, with 7.3 percent and 6.2 percent growth, respectively, between 1981 and 2002. The labor productivity of general

[10]China increased its share of the global production of electric machinery from 1.2 percent in 1981 to a staggering 16.0 percent in 2002.

Table 2.6 Malaysia's Share in World Production, 1981, 1990, and 2002

Sector	Share (%)		
	1981	1990	2002
Food products	0.4	0.3	0.9
Footwear, except rubber or plastic	0.1	0.0	0.2
Leather products	0.0	0.1	0.1
Textiles	0.2	0.2	0.5
Wearing apparel, except footwear	0.1	0.3	0.7
Wood products	0.5	0.5	1.1
Paper products	0.1	0.1	0.5
Chemicals	0.1	0.2	0.6
Petrochemicals	0.3	0.3	2.1
Plastic	0.2	0.2	1.0
Rubber	1.4	1.4	2.2
Glass	0.2	0.3	0.8
Nonferrous metals	0.5	0.2	0.6
Iron and steel	0.1	0.2	0.7
Fabricated metal products	0.2	0.2	0.4
Machinery, electric	0.3	0.8	2.4
Machinery, except electrical	0.1	0.1	1.1
Professional and scientific equipment	0.1	0.1	0.4
Transportation equipment	0.1	0.1	0.4
Others	0.1	0.2	0.5

Source: UNIDO INDSTAT3.

machinery increased 5.6 percent annually on average for the same period. Similarly, transportation equipment and electric machinery experienced labor productivity growth of around 5 percent (see table 2.7).

Reflecting the increase in value added per worker, the average wages in Malaysia increased by 3 percent per year on average between 1981 and 2002 (see table 2.8). In most subsectors, labor productivity growth is faster than wage growth. The food, footwear, leather, apparel, rubber, and nonferrous metals subsectors are the exceptions.

It is worrisome that the ratio of value added to output for the majority of the industrial subsector stagnated or declined between 1981 and 2002 (see table 2.9). Footwear, apparel, plastics, rubber, and nonferrous metals are the only subsectors in which value added has increased. In general machinery, the value added decreased from 36 percent of the output value to only 17 percent, the largest drop among the industrial subsectors. Similarly, the value-added ratio decreased in electric

Table 2.7 Value Added per Worker in Malaysia and Its Growth Rates, 1981, 1990, and 2002

Sector	Value added per worker (constant 2000 US$)			Average annual growth of value added per worker (%)		
	1981	1990	2002	1981–90	1990–2002	1981–2002
Food products	11,470	15,013	18,219	3.0	1.6	2.2
Footwear, except rubber or plastic	4,160	4,123	6,413	−0.1	3.7	2.1
Leather products	4,289	3,690	6,606	−1.7	5.0	2.1
Textiles	4,789	8,123	10,081	6.0	1.8	3.6
Wearing apparel, except footwear	3,115	4,488	5,828	4.1	2.2	3.0
Wood products	5,028	6,501	8,469	2.9	2.2	2.5
Paper products	7,403	12,216	16,460	5.7	2.5	3.9
Chemicals	12,516	45,766	54,788	15.5	1.5	7.3
Petrochemicals	131,786	113,745	466,388	−1.6	12.5	6.2
Plastic	4,686	7,531	15,257	5.4	6.1	5.8
Rubber	8,749	9,227	11,519	0.6	1.9	1.3
Glass	9,264	15,889	24,633	6.2	3.7	4.8
Nonferrous metals	18,006	13,536	20,956	−3.1	3.7	0.7
Iron and steel	8,928	21,635	24,868	10.3	1.2	5.0
Fabricated metal products	5,897	9,951	13,758	6.0	2.7	4.1
Machinery, electric	7,071	9,404	18,892	3.2	6.0	4.8
Machinery, except electrical	7,387	13,528	23,218	7.0	4.6	5.6
Professional and scientific equipment	5,280	6,859	13,899	2.9	6.1	4.7
Transportation equipment	9,726	20,320	28,608	8.5	2.9	5.3
Others	4,475	6,384	10,142	4.0	3.9	4.0

Source: UNIDO INDSTAT3.

machinery and transportation equipment. Among the East Asian economies, Malaysia's value-added ratios in machinery (both electric and nonelectric) are the lowest (see figure 2.2), even though since the early 1990s the Malaysian authorities—anticipating greater competition in the export of commoditized products—began introducing policies to induce the emergence of more knowledge-intensive activities and to increase the domestic content of traded items in the interests of growth over the longer term. These policies have attempted to enlarge the supply of skills as well as their quality; to attract more FDI into technology-intensive areas; to stimulate industrial links that lead to the entry of Malaysian firms producing higher-value intermediate and final products using advanced technologies; and overall, to upgrade the average quality of exports. However, the payoff from these policies has yet to materialize. The investment climate surveys conducted by the

Table 2.8 Average Wage in Malaysia and Its Growth Rates, 1981, 1990, and 2002

Sector	Average wage (constant 2000 US$)			Average annual growth of average wage (%)		
	1981	1990	2002	1981–90	1990–2002	1981–2002
Food products	2,301	3,267	4,261	4.0	2.2	3.0
Footwear, except rubber or plastic	1,775	2,100	3,560	1.9	4.5	3.4
Leather products	1,623	1,824	3,520	1.3	5.6	3.8
Textiles	2,125	2,749	3,710	2.9	2.5	2.7
Wearing apparel, except footwear	1,457	2,265	3,318	5.0	3.2	4.0
Wood products	2,303	2,642	3,041	1.5	1.2	1.3
Paper products	2,595	4,127	5,586	5.3	2.6	3.7
Chemicals	3,257	5,963	8,275	7.0	2.8	4.5
Petrochemicals	8,552	9,362	12,301	1.0	2.3	1.7
Plastic	1,718	2,317	4,364	3.4	5.4	4.5
Rubber	2,338	2,722	3,877	1.7	3.0	2.4
Glass	2,479	3,536	5,288	4.0	3.4	3.7
Nonferrous metals	3,493	3,945	6,105	1.4	3.7	2.7
Iron and steel	3,323	4,523	6,239	3.5	2.7	3.0
Fabricated metal products	2,317	3,357	4,855	4.2	3.1	3.6
Machinery, electric	2,234	2,915	5,168	3.0	4.9	4.1
Machinery, except electrical	2,542	3,687	5,780	4.2	3.8	4.0
Professional and scientific equipment	2,659	2,497	5,366	−0.7	6.6	3.4
Transportation equipment	3,345	3,833	6,127	1.5	4.0	2.9
Others	1,754	2,030	4,339	1.6	6.5	4.4

Source: UNIDO INDSTAT3.

World Bank in 2002 and 2008 indicate that real wages declined during this period, as did the average skill intensity of production, which would point to rising capital intensity and greater automation and down skilling of production activities. In effect, Malaysian industry appears to be sliding down the technological slope, and the incentives for workers to improve their skills are weakening.

This impression of an economy in which technological capabilities are relatively static (and may even be declining) and in which, largely as a consequence, industrial competitiveness is marking time is reaffirmed by indicators constructed by the World Economic Forum (WEF), the International Institute for Management Development (IMD), and the World Bank, among others, to benchmark performance. Malaysia is sandwiched between higher-income economies (Japan; the Republic of Korea; Singapore; and Taiwan, China) and lower-income economies (China, Indonesia, and Thailand) without much change in its ranking over the past five years. International

Table 2.9 Ratio of Value Added to Output Value, 1981, 1990, and 2002

Sector	1981 (%)	1990 (%)	2002 (%)
Food products	35	36	22
Footwear, except rubber or plastic	29	34	39
Leather products	30	25	31
Textiles	29	28	26
Wearing apparel, except footwear	35	32	34
Wood products	35	33	31
Paper products	39	40	38
Chemicals	30	38	33
Petrochemicals	20	25	22
Plastic	32	35	36
Rubber	23	27	27
Glass	46	49	46
Nonferrous metals	6	14	20
Iron and steel	23	20	23
Fabricated metal products	30	27	30
Machinery, electric	28	22	21
Machinery, except electrical	36	32	17
Professional and scientific equipment	36	29	26
Transportation equipment	36	29	30
Others	32	40	31

Source: UNIDO INDSTAT3.

Figure 2.2 Value-Added Ratios in Machinery in Selected East Asian Economies

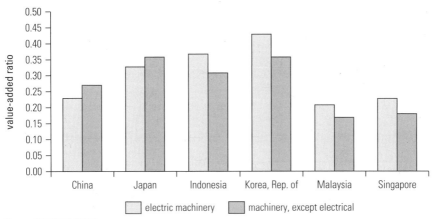

Source: UNIDO INDSTAT3.
Note: Data for China, Indonesia, and Singapore are from 2003; data for Japan, Korea, and Malaysia are from 2002.

Table 2.10 Rank Correlation among Selected Indexes on Technological Capabilities

Country	WEF	UNDP	ArCo	RAND	Rank mean	Standard deviation
Korea, Rep. of	7	5	15	16	10.8	5.56
Singapore	15	8	17	15	13.8	3.95
Malaysia	18	29	33	38	29.5	8.50
Thailand	31	36	37	41	36.3	4.11
Philippines	32	38	39	42	37.8	4.19
China	39	39	41	33	38.0	3.46
India	44	46	47	37	43.5	4.51
Indonesia	41	44	45	46	44.0	2.16

Source: Archibugi and Coco 2005.
Note: Countries are ordered by rank mean. Indexes are the WEF Technology Index, the United Nations Development Programme (UNDP) Technology Achievement Index, the ArCo Indicator of Technological Capabilities, the United Nations Industrial Development Organization (UNIDO) Industrial Development Scoreboard, and the RAND Corporation's Science and Technology Capacity Index. ArCo (developed by Archibugi and Coco) comprises patents granted in the United States, tertiary enrollment, and telecommunication infrastructure, among others. The RAND index contains many of the variables used in ArCo, as well as variables related to industrial performance, such as manufacturing value added, technology imports, and share of exports of high-tech goods. For the exact data and coverage of years for each indicator, see Archibugi and Coco (2005).

indicators of university rankings and international test scores in science and mathematics also indicate that Malaysia is situated in the middle of the pack and that its position has been stable over the years for which data are available (see the tables in annex 2.A). When one looks at a broader set of indicators, the picture becomes muddier, because each indicator places Malaysia and other countries at different positions, and correlations among indicators are low (Archibugi and Coco 2005). Technology-specific indicators rank Malaysia much lower, especially indicators that are not based on surveys but on structural characteristics (for example, the ArCo and RAND indicators in table 2.10). Although the subcomponents of the competitiveness indicators add detail, several fluctuate erratically from one year to the next without any apparent association to changes in policies. This fluctuation greatly dilutes the information content. The most that can be wrung out of these indicators is that the quality of human capital in Malaysia and its technological readiness lag the findings in the higher-income economies and that if demand for skills is softening, this lag could worsen. The estimates of TFP noted previously reinforce this impression. With the possible exception of electronic components, every major subsector lies well within the production and technology frontiers, with ample room for gains in productivity and the upgrading of technology. Hence, the potential for raising the contribution of TFP to GDP growth exists but is not being exploited.

Malaysia's labor force is expanding by close to 2 percent per year, and the average age of workers is 36 years. The rate of savings from 2001 to 2007 was equal to 33 percent of GDP (see table 2.11), far in excess of levels of domestic investment

Table 2.1.1 Macroindicators of Malaysia and Other Southeast Asian Economies, 1990–2007

Indicator	Japan	Singapore	China	Indonesia	Korea, Rep. of	Malaysia	Philippines	Taiwan, China	Thailand
GDP growth (%)									
1990–95	2.1	8.9	10.9	8.1	8.0	9.4	2.3	7.0	9.1
1996–2000	1.0	6.4	8.6	1.0	4.6	5.0	4.0	5.8	0.6
2001–07	1.5	5.3	10.2	5.1	4.7	4.8	5.0	3.8	5.1
Exports growth (%)									
1990–95	3.8	—	12.4	9.9	13.4	16.4	8.4	7.8	14.2
1996–2000	5.8	—	15.0	4.2	16.0	8.9	6.4	9.9	7.3
2001–07	6.1	—	23.5	7.6	11.2	5.7	6.0	7.8	6.3
Gross capital formation (% of GDP)									
1990–95	30.5	35.8	39.9	30.9	37.5	38.3	22.7	24.6	41.1
1996–2000	26.8	34.1	37.5	22.6	32.0	32.2	21.8	23.8	27.9
2001–07	23.4	20.9	41.1	24.1	29.7	22.4	16.3	20.2	27.0
Gross savings (% of GDP)									
1990–95	32.3	46.8	41.6	26.1	36.5	32.5	19.7	26.0	34.3
1996–2000	28.1	49.9	39.8	24.8	35.2	37.1	25.6	23.0	32.2
2001–07	26.5	44.9	45.6	27.0	31.5	32.5	31.3	25.6	28.7
Manufacturing (% of GDP)									
1990–95	—	27.0	33.3	22.3	27.2	25.8	24.1	28.6	28.7
1996–2000	22.7	25.1	32.5	26.2	27.6	29.7	22.2	24.7	31.4
2001–07	21.0	26.2	32.3	28.9	27.7	30.3	22.9	23.4	34.4

Source: World Bank World Development Indicators database.
Note: — = not available.

during the same period. The physical infrastructure is modern and well maintained, and Malaysia is not hamstrung by the power shortages besetting Indonesia, for example. The scope for raising TFP, the elasticity of factor supplies, the adequacy of the infrastructure, and a business environment appreciably superior to that of competitors such as China and India all suggest that Malaysia can move to a higher growth path commensurate with its level of income.[11] Malaysia's maintenance of growth rates of 6 percent per year or less, which was the average for Sub-Saharan Africa during 2006 and 2007, suggests that Malaysia was operating below potential from 2007 through 2008, as were several neighboring Southeast Asian economies.

Virtually all of the Southeast Asian economies have witnessed a marked slowdown from the growth rates achieved during 1990 to 1995. The exception is the Philippines, which has averaged growth rates from 2000 to 2007 that are double those from 1990 to 1995. Average annual export growth in all four countries has slowed, most strikingly in Malaysia but also in Thailand. Since the latter half of 2008, exports of all Southeast Asian economies have been dragged down by the global recession. Savings as a share of GDP remained reasonably stable in Indonesia, Malaysia, and Thailand, while they have risen in the Philippines to levels comparable with the others. Whether this stability will persist over the medium term is an open question. Quite remarkably, the share of manufacturing has climbed in Malaysia, Indonesia, and Thailand and has remained broadly the same in the Philippines. Once again, the dramatic weakening of export demand starting in 2008 raises questions about future trends in manufacturing and its role as a leading sector.

Growth rates of 5 percent to 6 percent from 2000 to 2008 were respectable, and these rates—or even higher levels—might well be achievable over the medium term once the global economy recovers. However, if Malaysia and other Southeast Asian countries are to climb back to a high-growth path, they will need to craft a fresh strategy that draws residual momentum from a first phase of industrialization now coming to a close. Now is the time for a new strategy that recognizes the harsher international trading environment, that firmly embraces the necessity for institutional changes and supports vigorous entrepreneurial initiatives, and that emphasizes innovation to take the economy into more promising industrial waters.

---

[11]However, the railway system needs an infusion of capital.

Annex 2.A Indicators of Competitiveness

Competitiveness

IMD's ranking of Malaysia's global competitiveness changed little between 2004 and 2008 (see table 2.A.1), although it fluctuated in 2004 and 2005. China's ranking, however, has improved from 22 in 2004 to 17 in 2008, surpassing all the economies in East Asia except for Singapore and Taiwan, China.

When one looks at the subcomponents of Malaysia's aggregate competitiveness indicator, two factors stand out. First, between 2004 and 2008, there was virtually no change to the infrastructure component, which consistently ranked lowest among the four subcomponents used by IMD and which was a drag on competitiveness (see table 2.A.2). Second, government efficiency deteriorated from 2004. Drilling down into the performance of components of infrastructure in the past five years, one finds that total health expenditure, energy intensity, dependency

Table 2.A.1 IMD Global Competitiveness Ranking, 2004–08

Economy	2004	2005	2006	2007	2008
Singapore	2	3	3	2	2
Taiwan, China	12	11	17	18	13
China	22	29	18	15	17
Malaysia	16	26	22	23	19
Thailand	26	25	29	33	27
Korea, Rep. of	31	27	32	29	31
Philippines	43	40	42	45	40
Indonesia	49	50	52	54	51
Japan	21	19	16	24	22

Source: IMD 2008.
Note: Economies are ordered by rank in 2008.

Table 2.A.2 Subcomponents to IMD Global Competitiveness Rankings: Malaysia, 2003–07

Subcomponent	2003	2004	2005	2006	2007
Overall performance	21	16	26	22	23
Economic performance	23	15	8	10	12
Government efficiency	14	16	23	19	21
Business efficiency	18	12	22	19	15
Infrastructure	27	25	28	27	26

Source: IMD 2008.

ratio, total R&D spending, and R&D personnel per capita appear as the weakest components (IMD 2003, 2004, 2005, 2006, 2007). In 2007, the indexes of R&D are replaced by two educational measures: total public expenditure on education and secondary school enrollment (IMD 2007). In the government efficiency subcategory, IMD identifies government subsidies to private and public companies, government budget surplus and deficit, foreign investors, exchange rate stability, price control, and discrimination as recurring weaknesses. The indicator for foreign investor refers to the freedom—or lack thereof—of foreign investors to acquire controlling stakes in local firms (IMD 2003, 2004, 2005, 2006, 2007). In Indonesia, economic performance and infrastructure declined between 2003 and 2007, the former more than the latter. Government efficiency remained more or less stable, while business efficiency improved a little. In the infrastructure sector, medical assistance and total public expenditure continued to be the weakest performing areas. In Thailand, rankings of government and business efficiency, as well as infrastructure, deteriorated between 2003 and 2007. Within government efficiency, the foreign investors component was consistently one of the weakest until 2006. Within infrastructure, the poorest-performing segments were medical assistance, total health expenditure, and total expenditure on R&D.

Malaysia's ranking in the global competitiveness index computed by the World Economic Forum declined from 2006 to 2007, but by only 2 points (see table 2.A.3). However, the WEF's index shows that Malaysia ranks in the middle among Asian economies.

A closer look at the subcategories reveals which factors fall below Malaysia's overall ranking. These subcategories are macroeconomy, technological readiness, market size, higher education and training, and health and primary education (see table 2.A.4). A juxtaposition of the lowest-rated components from the two

Table 2.A.3 WEF Global Competitiveness Index, 2006–07 and 2007–08

Economy	2006–07	2007–08
Singapore	8	7
Japan	5	8
Korea, Rep. of	23	11
Taiwan, China	13	14
Malaysia	19	21
Thailand	28	28
China	35	34
Indonesia	54	51
Philippines	75	67

Sources: Lopez-Claros and others 2006; Porter, Schwab, and Sala-i-Martin 2007.
Note: Economies are ordered by rank in 2007–08.

Table 2.A.4 WEF Global Competitiveness Rankings of Malaysia and Its Major Competitors, 2007–08

Index components	Malaysia	Thailand	Korea, Rep. of	China	Indonesia	Singapore	Philippines	Japan	Taiwan, China
Global competitiveness index	21	28	11	34	54	7	71	8	14
Basic requirements subindex	21	40	14	44	82	3	93	22	19
Institutions	20	47	26	77	63	3	95	24	37
Infrastructure	23	27	16	52	91	3	94	9	20
Macroeconomy	45	30	8	7	89	24	77	97	26
Health and primary education	26	63	27	61	78	19	86	23	6
Efficiency enhancers subindex	24	29	12	45	37	6	60	13	17
Higher education and training	27	44	6	78	65	16	62	22	4
Goods market efficiency	20	34	16	58	23	2	64	19	17
Labor market efficiency	16	11	24	55	31	2	100	10	22
Financial market sophistication	19	52	27	118	50	3	77	36	58
Technological readiness	30	45	7	73	75	12	69	20	15
Market size	29	17	11	2	15	50	24	4	16
Innovation factors subindex	19	39	7	50	34	13	65	2	10
Business sophistication	18	40	9	57	33	16	55	3	14
Innovation	21	36	8	38	41	11	79	4	9

Sources: Lopez-Claros and others 2006; Porter, Schwab, and Sala-i-Martin 2007.

competitiveness indexes suggests that education at all levels and associated technological readiness are compromising Malaysia's competitiveness with Korea and Singapore. For Thailand, health and primary education ranked the lowest, followed by financial market sophistication, institutions, and technological readiness, in that order. Infrastructure, macroeconomy, health and primary education, and technological readiness were the weakest areas for Indonesia. These weaknesses are not too dissimilar to those of Malaysia.

Business Environment

Complementing the indicators of competitiveness is a number of indexes devised to assess the general business environment. One is the Doing Business indicator constructed by the World Bank. Among East Asian economies, the business environment facing Malaysian firms is relatively favorable (see table 2.A.5). Between 2005 and 2008, there were few changes. Korea dropped below Malaysia. China, India, and Indonesia greatly improved their scores. However, private businesses in those countries still face more adverse conditions than do businesses elsewhere.

Among the 10 components from which the aggregate Doing Business indicator is constructed, only 2 components rank higher than the overall indicator for Malaysia (see table 2.A.6). Malaysia does well in providing credit to new firms and in protecting investors. However, starting a business is cumbersome, despite ready access to credit for established firms with adequate collateral. The lowest-rated component is dealing with licenses. Although the cost of compliance to license in Malaysia is marginally larger than that for member countries of the Organisation for Economic Co-operation and Development (OECD), the cost in

Table 2.A.5 Doing Business Indicators, 2006–09

Economy	2006	2007	2008	2009
Singapore	2	1	1	1
Japan	10	11	12	12
Thailand	20	18	15	13
Malaysia	21	25	24	20
Korea, Rep. of	27	23	30	23
Taiwan, China	35	47	50	61
China	91	93	83	83
Indonesia	115	135	123	129
Philippines	113	126	133	140

Sources: World Bank 2005a, 2006b, 2007b, 2008b.
Note: Economies are ordered by rank in 2009.

Table 2.A.6 Individual Components of Doing Business Rankings: Malaysia, 2005–08

Ease of ...	2005 rank	2006 rank	2007 rank	2008 rank
Doing business	25	25	24	20
Starting a business	66	71	74	75
Dealing with licenses	134	137	105	104
Employing workers	37	38	43	48
Registering property	68	66	67	81
Getting credit	3	3	3	1
Protecting investors	3	4	4	4
Paying taxes	49	49	56	21
Trading across borders	41	46	21	29
Enforcing contracts	78	81	63	59
Closing a business	47	51	54	54

Sources: World Bank 2005a, 2006b, 2007b, 2008b.
Note: The 2005 rankings have been recalculated to reflect changes to the 2006 methodology and the addition of 20 new countries.

terms of number of procedures (25 in Malaysia versus 15 in OECD countries) and the number of days (261 days versus 154 days) are close to double OECD averages. Nonetheless, in terms of its ranking for dealing with licenses, Malaysia improved its position in 2008. Other areas in which Malaysia saw improvement were trading across borders and enforcing contracts.

3

Analyzing Comparative Advantage and Industrial Change
Reading the Export Trade Tea Leaves

Malaysia's future economic performance will be strongly influenced by several factors. Among them, the growth and composition of exports supported by investments that increase the domestic value added by firms will continue to play a leading role. Malaysia is an unusually open economy.[1] Total trade (exports plus imports of goods and services) amounted to 210 percent of gross domestic product (GDP), which is higher than the ratio for other East Asian economies, with the exception of Singapore (see table 3.1).[2]

Export volume and domestic value added strongly affect the tempo of economic activity and per capita income. Although export earnings came under great pressure during the second half of 2008 and declined in 2009, a revival will be a function of increased market share for existing products, product diversification, higher unit values of exports, and penetration into new markets. Such a revival will be easier if the expansion of international trade quickly returns to earlier trend rates. It could be far more challenging if global recovery is slow and import demand from the United States and the European Union (EU) remains weak.

Generally, those firms that are larger and more skill intensive and that have achieved above-average productivity levels venture beyond the national market and

---

[1] On the research relating productivity and growth to openness, see Edwards (1998). Openness also promotes innovation, according to a study of Organisation for Economic Co-operation and Development countries (Hulten and Isaksson 2007). Malaysia's high ratio of trade to gross domestic product reflects the importance of processing, assembly, and testing activities in the electronics subsector and low domestic value added.

[2] A part of the explanation for Malaysia's high ratio of trade to GDP is the level of foreign direct investment in the tradable sector (Sjöholm 1999).

Table 3.1 Exports and Imports of Goods and Services as a Percentage of GDP, 2007

Economy	Percentage of GDP
Singapore	463.0
Malaysia	210.0
Taiwan, China	140.0
Thailand	132.5
Philippines	83.3
Korea, Rep. of	90.4
China[a]	72.0
Indonesia	54.7
Japan[b]	27.3

Source: World Bank World Development Indicators database.
a. Data are for 2006.
b. Data are for 2005.

are likely to spearhead recovery and future growth. And the intention to expand by exporting frequently motivates investment (in equipment, training, and research and development, for example); organizational changes; and innovation by firms (see Alvarez 2007; Aw, Roberts, and Winston 2005; Bernard and Jensen 1999; Bernard and others 2007; Clerides, Lach, and Tybout 1998; De Loecker 2007; Hallward-Driemeier, Iarossi, and Sokoloff 2002; Levinsohn and Petrin 1999).[3] Investment in both equipment and technology development plays a big part in enhancing efficiency and upgrading products and production processes. In turn, exporting stimulates growth. In large economies, domestic market demand can serve as a buffer against external shocks. It can also be a source of competitive pressures, which induce firms to be efficient and to innovate. In smaller economies, however, the odds against that happening are great because the domestic market is a very partial substitute for exports (and can constrain firms' ability to achieve scale economies). For this reason, growth in the more trade-dependent economies plummeted rapidly during the final quarter of 2008. In today's integrating global economy, dynamic firms must be prepared to test the waters of the export market. Their success under a broadly favorable trading environment can set the stage for a virtuous spiral.

Exports and Industrial Change

In 2007, Malaysia exported US$202.4 billion worth of goods and services—well over twice the value of exports in 1995. During the last half of the 1990s, Malaysia's export

---

[3] Roberts and Tybout (1997) describe a process of turnover whereby less productive and less skill-intensive firms exit, thus raising average productivity. More productive firms are more skill and capital intensive and are more likely to be in industries where trade costs are falling (Bernard and Jensen 2001; Bernard and others 2007; Bernard, Jensen, and Schott 2003).

Table 3.2 Exports of Goods and Services, 2000–07

Economy	Value (US$ billion)			Annual growth rate (%), 1995–2000	Annual growth rate (%), 2000–07
	1995	2000	2007		
China[a]	168.0	279.6	1,061.7	8.9	21.0
Indonesia	53.2	67.6	127.1	4.1	8.2
Korea, Rep. of	149.1	208.9	442.2	5.8	9.8
Malaysia	83.6	112.4	202.4	5.1	7.6
Philippines	26.9	42.1	56.7	7.7	3.8
Taiwan, China	129.3	172.6	282.7	4.9	6.4
Thailand	70.3	82.0	167.3	2.6	9.3
Japan[b]	480.9	512.7	652.4	1.1	4.1
Singapore	—	164.0	372.6	—	12.4

Source: World Bank World Development Indicators database.
Note: — = not available.
a. Data are for 2006.
b. Data are for 2005.

growth was 5.1 percent per year and was exceeded only by China, the Republic of Korea, and the Philippines (see table 3.2). Between 2000 and 2007, Malaysia's export growth rose to 7.6 percent per year, somewhat higher than that of Taiwan, China, and the Philippines. Meanwhile, the annual increase in exports from Thailand rose to 9.3 percent and those from China accelerated to 21.0 percent per year. The share of Malaysia's exports in the world market declined steadily from 1995 to 2007, falling below 1.4 percent in 2007 (see figure 3.1). Furthermore, the share of Malaysia's exports to the United States is trending downward (see figure 3.2). Although Malaysia's exports to China declined between 2002 and 2005, this trend reversed between 2005 and mid-2008 (see figure 3.3).[4] Similar trends can be seen in exports by other Southeast Asian countries. Exports of the three other Southeast Asian countries to the United States are also on a downward trend, but their exports to China have been rising. All of the Southeast Asian countries saw their exports plummet starting in the last quarter of 2008.

A glance at figures 3.4 and 3.5 shows that from 1995 to 2007, in value terms, Malaysia's fastest-growing exports by broad category were electrical and electronic products and engineering products, followed by primary products (mainly palm oil and derivatives). Processed exports and other low-technology manufactures increased more gradually in the aggregate, although individual items in these categories grew much faster. Automotive and other high-technology exports remained flat over the entire 12-year period. With domestic value added in the

---

[4] An examination of the top 20 fastest-growing (by value) export commodities worldwide between 2000 and 2007 reveals that Malaysia did not export any. Many of them were metallic and chemical compounds or food products.

Figure 3.1 Share of Malaysia's and Selected Southeast Asian Countries' Exports to the World, 1995–2007

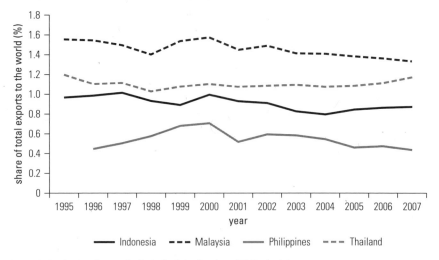

Source: United Nations Commodity Trade Statistics Database (UN Comtrade).

Figure 3.2 Share of Malaysia's and Selected Southeast Asian Countries' Exports to the United States, 1995–2007

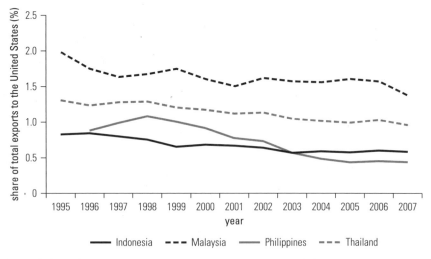

Source: UN Comtrade.

Figure 3.3 Share of Malaysia's and Selected Southeast Asian Countries' Exports to China, 1995–2007

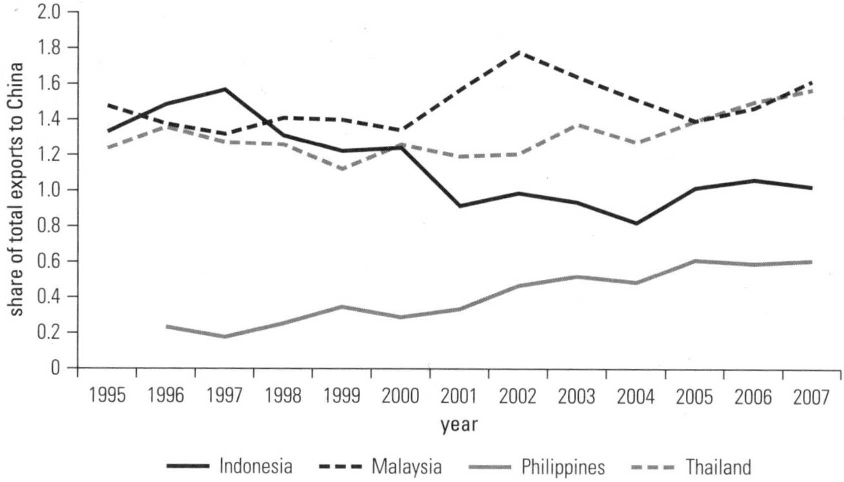

Source: UN Comtrade.

Figure 3.4 Exports of Malaysia by Type of Manufacture, 1995–2007

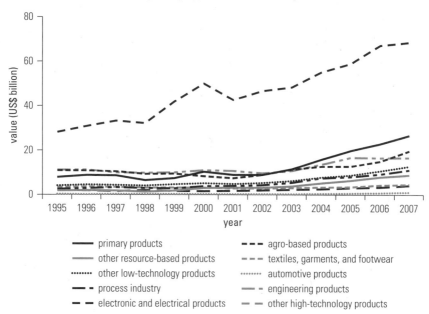

Source: UN Comtrade.

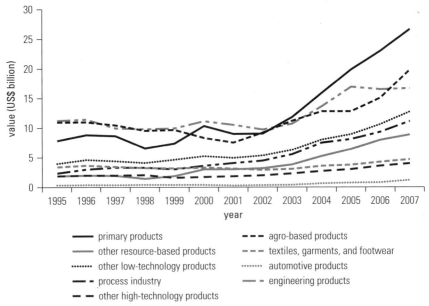

Figure 3.5 Exports of Malaysia by Type of Manufacture, Excluding Electronics, 1995–2007

Source: UN Comtrade.

merchandise export sector remaining constant at approximately 30 percent, the picture that emerges is of a widening of processing and assembly activities in the electronics subsector,[5] no increase in the export shares of other medium- or higher-technology industries such as auto parts, and expanding exports of palm oil and derivatives of the palm oil industry.[6] Figure 3.6 shows that Malaysia's electronic and electrical exports were higher than four other neighboring countries—namely, India, Indonesia, the Philippines, and Thailand.

---

[5] By 1980, integrated circuits represented 8.4 percent of exports from Malaysia. The share increased to 16.3 percent by 1993. Electronic and electrical equipment accounted for 58 percent of exports in 2005 (Reinhardt 2000).

[6] Although there is scope for adding value to the exports of natural resource–intensive products, it is not happening with such exports from Malaysia and Thailand. For instance, the material costs of soap exported by Malaysia alone account for 75 percent of production costs. Similarly, for petrochemicals, materials account for as much as 90 percent of the costs. Even in rubber products, Malaysia's exports tend to be from the lower end of the spectrum, such as latex gloves and swimming caps, for which raw material is the most significant component. Both Malaysia and Thailand have found it hard to move up to higher value-added items that are based on domestic natural resources (Reinhardt 2000).

Figure 3.6 Exports of Electronic and Electrical Manufactures, 1995–2007

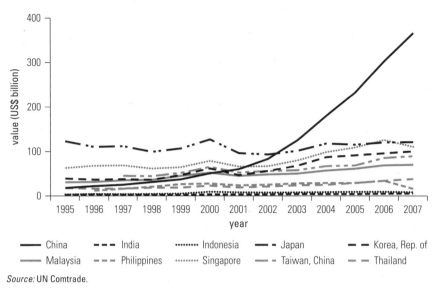

Source: UN Comtrade.

The notable feature of Malaysia's fastest-growing exports during 2000 to 2007, at a six-digit level of disaggregation, was their diversity (see table 3.3) and the low prevalence of high-technology electronic products (only 1 item in the top 20) vis-à-vis resource-based and processed commodities, especially petrochemical products.

In table 3.4 and figure 3.7, four trends stand out. First, from 1995 to 2007, the shares of electronics in total merchandise exports changed little among middle-income East Asian economies. The declines in the cases of Japan; Korea; and Taiwan, China, are largely explained by the shift in production from those economies to China mediated by foreign direct investment. Rising production costs also induced multinational corporations (MNCs) to shift their labor-intensive activities from Singapore to China. This shift explains the massive jump in China's exports of electronics. In 1995, they represented less than 10 percent of total exports. Twelve years later, the share of electronics had climbed to 30.1 percent, whereas the total value of China's exports rose from about US$148 billion to US$1.2 trillion in 2007. By 2005, China had become the largest exporter of electronics in the world, edging out the United States.

A second trend is the drop in the shares of textiles and low-tech manufactures in the exports from East Asian economies (see table 3.4). In contrast, combined exports of primary, agro-based, and resource-based products have increased in Indonesia, Korea, Malaysia, and the Philippines. Curiously, the share of primary product exports from Thailand declined during this period. A fourth and potentially important trend is the increasing share of engineering and automotive products in the exports from China, Korea, and Thailand. By contrast, engineering products as a share of Malaysia's exports fell from 15.4 percent to 9.6 percent.

Table 3.3 Fastest-Growing Exports from Malaysia by Level of Technology, 2000–07

Product	Product name	Annual growth rate (2000–07)	Technology class
900921	Photocopying equipment with an optical system, not elsewhere specified	322.6	HT1
290124	Buta-1, 3-diene, and isoprene	267.2	RB2
560420	High-tenacity manmade yarn, rubber, plastic coated or impregnated	202.2	LT1
720712	Semifinished bars, iron or nonalloy steel, <0.25% carbon, rectangular, not elsewhere specified	198.7	MT2
721912	Hot-rolled stainless steel coil, width of >600 mm, thickness of 4.75–10 mm	198.5	MT2
720720	Semifinished product, iron or nonalloy steel, >0.25% carbon	195.9	MT2
240391	Homogenized or reconstituted tobacco	194.2	RB1
100640	Rice, broken	188.4	PP
230890	Vegetable wastes and residues, not elsewhere specified, for animal feed	186.4	PP
842382	Weighing machinery having a capacity of 30–5,000 kg	164.2	MT3
470693	Semichemical pulps of other fibrous material	163.9	RB1
271312	Petroleum coke, calcined	163.8	RB2
570291	Carpets of wool or fine hair, woven, made up, not elsewhere specified	160.1	LT1
848250	Bearings, cylindrical roller, not elsewhere specified	155	MT3
890190	Cargo vessels other than tanker or refrigerated	154.8	MT3
480451	Paper, kraft, >225 g/m^2, unbleached, uncoated, not elsewhere specified	150.1	RB1
681250	Asbestos clothing, accessories, footwear, and headwear	149	RB2
871494	Bicycle brakes, parts thereof	147.1	MT1
722692	Cold-rolled alloy-steel not elsewhere specified, not further worked, <600 mm wide	146.4	LT2
480210	Paper, handmade, uncoated	146.3	RB1

Source: UN Comtrade.
Note: HT1 = electronic and electrical products; LT1 = textiles, garments, and footwear; LT2 = other low-technology products; MT1 = automotive products; MT2 = process industry; MT3 = engineering products; PP = primary products; RB1 = agro-based products; RB2 = other resource-based products. Technology classification is based on Lall (2000).

Export Overlap

How the changing composition of Malaysia's exports has affected the degree of competitive pressure on its exporters is brought into a sharper focus by tables 3.5 and 3.6 and figures 3.8 and 3.9, which present the percentage of overlap between Malaysia's exports and those of the comparator countries with reference to items at

Table 3.4 Leading Export Sectors, 1995 and 2007

Economy	Percentage of total exports	
	1995	2007
China		
Electronic and electrical products	10.8	30.1
Textiles, garments, and footwear	30.9	16.5
Engineering	10.6	15.3
Other low-technology products	15.8	14.8
Indonesia		
Primary products	36.0	35.4
Agro-based products	18.6	17.6
Other resource-based products	8.3	12.5
Textiles, garments, and footwear	15.9	9.4
Japan		
Engineering products	26.5	25.7
Automotive products	18.0	23.3
Electronic and electrical products	28.1	17.6
Process industry	7.1	9.0
Korea, Rep. of		
Electronic and electrical products	29.5	26.9
Engineering products	15.1	18.3
Automotive products	7.4	13.2
Other resource-based products	4.4	10.5
Malaysia		
Electronic and electrical products	38.7	39.3
Primary products	10.9	15.3
Agro-based products	15.2	11.4
Engineering products	15.4	9.6
Philippines[a]		
Electronic and electrical products	49.0	47.9
Primary products	7.3	9.9
Agro-based products	7.3	9.0
Other resource-based products	3.0	7.3
Taiwan, China[b]		
Electronic and electrical products	36.0	35.5
Other low-technology products	16.1	13.5
Process industry	8.6	12.0
Engineering products	13.7	11.3
Thailand		
Electronic and electrical products	23.0	23.2
Engineering products	10.1	13.6
Primary products	16.9	11.4
Other low-technology products	11.1	10.2

Source: UN Comtrade.
a. Data for initial year are from 1996.
b. Data for initial year are from 1997.

Figure 3.7 Composition of Exports by Type of Manufactures, 1995 and 2007

- electronic and electrical products
- other high-technology products
- automotive products
- process industry
- engineering products
- textiles, garments, and footwear
- other low-technology products
- agro-based products
- other resource-based products
- primary products

Source: UN Comtrade.

Table 3.5 Share of Overlapping Commodities, 1995, 2000, and 2007

Economy	Share (%)		
	1995	2000	2007
China	95.7	98.1	93.8
Indonesia	76.6	90.1	88.1
Japan	94.4	95.3	91.5
Korea, Rep. of	92.2	92.3	89.1
Philippines	54.5[a]	57.6	58.3
Singapore	98.0	97.7	91.7
Taiwan, China	93.6[b]	92.1	94.4
Thailand	89.7	91.6	91.8

Source: UN Comtrade.
a. Data are from 1996.
b. Data are from 1997.

Table 3.6 Share of Overlapping Trade Values, 1995, 2000, and 2007

Economy	Share (%)		
	1995	2000	2007
China	39.2	52.2	65.0
Indonesia	22.5	27.4	29.7
Japan	51.6	67.6	46.3
Korea, Rep. of	37.0	49.8	44.9
Philippines	16.0[a]	18.2	10.9
Singapore	56.2	58.9	38.6
Taiwan, China	46.5[b]	52.8	46.1
Thailand	28.2	34.4	35.9

Source: UN Comtrade.
a. Data are from 1996.
b. Data are from 1997.

a six-digit level of disaggregation.[7] Malaysia's exports overlap the most with Taiwan, China (94.4%), and China (93.8 percent), followed by Thailand, Singapore, and Japan, with Philippine exports overlapping the least. When the overlap is viewed

---

[7] The pairwise share of overlapping commodities is calculated as the number of products exported by Malaysia and each one of its competitors (for example, Indonesia) in a certain year, expressed as a proportion of the total number of commodities exported by Malaysia in that same year.

Figure 3.8 Share of Overlapping Commodities, 1995, 2000, and 2007

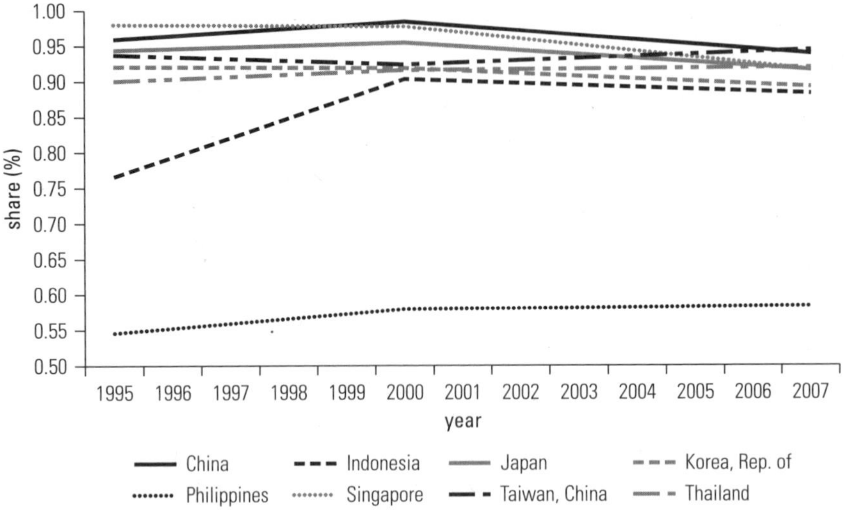

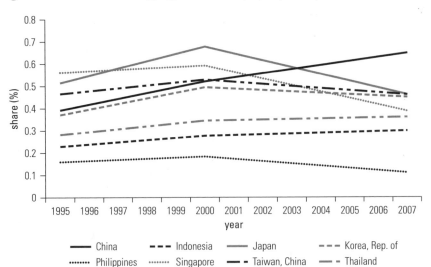

Source: UN Comtrade.
Note: Data for the Philippines begin in 1996, and data for Taiwan, China, begin in 1997.

Figure 3.9 Share of Overlapping Trade Values, 1995, 2000, and 2007

Source: UN Comtrade.
Note: Data for the Philippines begin in 1996, and data for Taiwan, China, begin in 1997.

after adjusting for the value of the common exports,[8] Malaysia's closest competitor is China, followed by Japan; Taiwan, China; and Korea, in that order, mainly because all have large exports of electronic equipment and components (see table 3.6).

The extent of overlap in value terms following this adjustment ranges from 65.0 percent for China to 10.9 percent for the Philippines. By this measure, Malaysia's export mix overlaps the least with the Philippines' even though both export a variety of electronic components and assembled products.

From this information, it appears that Malaysia's export growth in dollar terms during 2000 to 2007 was restrained because of an increasing concentration on electronics and rapidly intensifying competition from China, whose own production and trade in electronics have risen dramatically.[9]

To assess the implications of China's exports for the Southeast Asian exports more directly, one may use the methodology of Greenaway, Mahabir, and Milner (2008) and estimate a modified gravity model that explains trade flows with reference to the size of an economy and the distance among trading nations. The dependent variable is the log of the imports of country i from Association of Southeast Asian Nations (ASEAN) country j from 1990 to 2007, the latest year for which we have data. In addition to the standard set of variables for gravity models,[10] the specification adds the log of China's exports and two period dummies, one for 1990 to 1996 and one for 2001 to 2007.[11] These periods can be classified as the precrisis period (1990–96), the postcrisis period (1997–2000), and the period after China's accession to the World Trade Organization (2001–07). The results are shown in table 3.7. In all specifications, the standard variables for the gravity model have the expected signs, and they are all significant, except for the coefficient

---

[8]The overlapping trade value for a commodity X exported by Malaysia and Indonesia, for example, is calculated by identifying the minimum of the export values of X for Malaysia and Indonesia. The pairwise share of trade values for the overlapping commodities is then calculated by summing all the overlapping products and expressing it as a proportion of the total trade value of Malaysia's exports.

[9]Although competition from China stiffened between 1996 and 2003, China's rising exports slowed those of Southeast Asian economies to a limited extent only (Hanson and Robertson 2007).

[10]These variables are the GDPs of the importing and exporting countries, per capita GDPs of importing and exporting countries, distance between two countries, land areas, dummies for landlocked countries, dummies for island countries, dummy for sharing the border, dummy for having a common language, dummy for having a colonization relationship, dummy for having common colonizers, and indicator for country risk and corruption. The GDP data come from the World Bank World Development Indicators database. The trade statistics are from the Direction of Trade data maintained by the International Monetary Fund. Distance and other country characteristics are compiled by Andrew Rose and are available at http://faculty.haas.berkeley.edu/arose/datastataspiegel.zip.

[11]The same specifications were run for each period separately. The results are similar to those reported in table 3.7.

Table 3.7 Full Time Period, 1990–2006, with Subperiod Dummies

	In imports of country *i* from Asian country *j*		
	OLS	IV	IV, GMM
ln China's exports to country *i*	0.673 (33.10)**	−0.914 (6.34)**	−0.932 (7.25)**
Log of GDP of importing country	0.421 (18.74)**	1.737 (14.32)**	1.743 (15.79)**
Log of per capita GDP of importing country	0.024 (1.08)	−0.148 (4.39)**	−0.123 (3.38)**
Log of GDP of exporting country	2.816 (40.97)**	2.156 (19.64)**	2.185 (20.67)**
Log of per capita GDP of exporting country	1.126 (28.65)**	0.624 (8.99)**	0.634 (9.49)**
Log of distance	−0.794 (20.07)**	−1.930 (16.85)**	−1.946 (19.45)**
Log of product of land areas	−0.102 (8.61)**	−0.172 (10.03)**	−0.173 (9.83)**
Number landlocked 0/1/2	−0.815 (16.73)**	−1.982 (16.10)**	−1.973 (16.42)**
Number islands 0/1/2	−0.362 (7.27)**	−0.954 (11.17)**	−0.903 (11.47)**
Land border dummy	1.536 (4.71)**	−0.738 (1.52)	−0.761 (3.51)**
1 for common language	0.075 (1.66)+	0.259 (4.13)**	0.247 (3.86)**
Dummy for common colonizer post 1945	0.538 (7.58)**	0.803 (8.15)**	0.872 (8.60)**
Dummy for pairs ever in colonial relationship	1.117 (5.93)**	2.387 (8.58)**	2.431 (13.51)**
International Country Risk Guide	0.008 (3.38)**	0.011 (3.33)**	0.008 (1.98)*
Subperiod = 1990–96	0.390 (8.51)**	0.664 (9.98)**	0.633 (9.77)**
Subperiod = 2001–07	−0.315 (7.15)**	−0.557 (8.82)**	−0.574 (9.37)**
Constant	−79.250 (38.36)**	−54.242 (15.20)**	−54.699 (16.16)**
Observations	7,456	7,456	7,456
R-squared	0.79	0.62	

Source: Authors' calculations.
Note: GMM = general method of moments; IV = instrumental variables; OLS = ordinary least squares; + = significant at the 10 percent level; * = significant at the 5 percent level; ** = significant at the 1 percent level. Absolute value of *t*-statistics is in parentheses.

estimate on the per capita GDP of the importing country. The key variable is the coefficient estimate on China's exports. The estimate in ordinary least squares (OLS) is positive and significant. However, because endogeneity is a concern, OLS is not an appropriate specification. It is provided as a benchmark only. The preferred method is to use instrumental variables. The coefficient estimates on China's exports using instrumental variables are negative and significant. This finding implies that an increase in China's exports leads to a reduction in exports from ASEAN countries. The period dummies also show the dynamics of this change. The coefficient estimate for the period from 1990 to 1996 is positive. Hence, the effect of China's exports on other countries was smaller relative to the 1997 to 2000 period. The coefficient estimate on the 2001 to 2007 period is negative, thus implying that the effect of China's exports on other countries has increased. These results confirm the general observation that the rise of China's exports is intensifying the competitive pressures in the global market on Southeast Asian economies in particular.

The analysis examines only the overall bilateral trade flow between two countries. To gauge the changing competitiveness of Malaysian exports relative to those of China, one needs to examine trends in bilateral trades at the commodity level. Dynamic revealed competitiveness (DRC) is one such measure (Gallagher, Moreno-Brid, and Porzecanski 2008). It is based on the changing market (import) shares of a commodity i between two time periods. Here, the main interest is the changing import share of Malaysian products in three important markets: the EU15,[12] Japan, and the United States. They represent the major markets for Malaysian exports. A positive DRC means that the share of Malaysian products has increased in the importing country or region. Furthermore, by combining the DRC measures for China and Malaysia, it is possible to determine which products from two different countries are in direct competition. For instance, if the DRC is negative for a commodity exported by Malaysia but is positive for China, then the commodity is said to be in "direct threat." However, if the DRC is positive for both countries, then it is in "partial threat." Table 3.8 lists the proportion of Malaysian exports that are under direct threat and partial threat from China according to six-digit commodity-level trade data. Close to one-third of the products exported by Malaysia to those three important markets are directly threatened by products from China. When those products are combined with those under partial threat, 50 to 60 percent of Malaysian exports are in direct competition with those from China. The intensifying competition is most visible in the European market, even though overall competition is increasing in all three markets.

Table 3.9 lists Malaysia's top 10 exports to the U.S. market by trade value, excluding petroleum. Except for the nonsurgical rubber gloves, all other products are electrical and electronic components, reflecting the overall structure of

---

[12]The EU15 are the 15 members of the EU prior to its expansion on May 1, 2004: Austria, Belgium, Denmark, Finland, France, Germany, Greece, Ireland, Italy, Luxembourg, the Netherlands, Portugal, Spain, Sweden, and the United Kingdom.

Table 3.8 Degree of Competition between Malaysian and Chinese Exports, 1990–2007

Country or region	Degree of competition (%)		
	1990–91	2000–01	2006–07
EU15			
Direct threat	14.4	26.6	30.0
Partial threat	16.5	21.0	28.5
Total	30.8	47.5	58.5
Japan			
Direct threat	20.4	33.4	28.2
Partial threat	25.6	18.7	20.9
Total	46.0	52.1	49.2
United States			
Direct threat	25.6[a]	28.4	28.9
Partial threat	23.5[a]	22.4	26.0
Total	49.1[a]	50.8	54.9

Source: Authors' calculations using data from UN Comtrade.
a. Data are from 1991–92.

Table 3.9 Dynamic Revealed Competitiveness Position for Malaysia's Top 10 Nonoil Exports to the United States, 2007

Product	Product name	Share of U.S. imports		DRCP	
		1991	2007	Malaysia	China
847120	Digital computers with central processing and input-output units	0.0	28.4	28.4	61.6
847330	Parts and accessories of data-processing equipment, not elsewhere specified	1.1	14.4	13.3	44.8
851730	Telephonic or telegraphic switching apparatus	0.0	15.6	15.5	32.5
852520	Transmit-receive apparatus for radio, television, and so forth	8.8	4.7	–4.0	31.1
854219	Monolithic integrated circuits, except digital	12.8	15.3	2.5	3.4
854211	Monolithic integrated circuits, digital	11.8	9.0	–2.8	9.2
851790	Parts of line telephone or telegraph equipment, not elsewhere specified	1.8	12.8	11.0	19.3
847193	Computer data storage units	2.7	6.5	3.8	29.4
401519	Rubber gloves, other than surgical	45.1	48.0	2.9	8.2
852530	Television cameras	0.0	5.0	4.9	41.8

Source: Authors' calculations using data from UN Comtrade.

Malaysian exports. Among these top 10 products, Malaysia increased its market share in 8 of them, in some cases quite markedly, such as computers. For those where Malaysia lost market share, China increased its market share. Therefore, those products are under direct threat from China. Even for the eight products where Malaysia increased its market share, however, the products are under partial threat, because China's increase in market share was larger. Among the

Table 3.10 Dynamic Revealed Competitiveness Position for Malaysia's Top 10 Nonoil Exports to Japan, 2007

Product	Product name	Share of Japanese imports 1990	Share of Japanese imports 2007	DRCP Malaysia	DRCP China
441211	Plywood 1 or 2 outer ply tropical hardwood, ply <6 mm	0.7	56.6	55.9	0.9
854211	Monolithic integrated circuits, digital	3.6	4.0	0.4	8.3
851790	Parts of line telephone or telegraph equipment, not elsewhere specified	0.8	7.6	6.9	46.8
854219	Monolithic integrated circuits, except digital	0.4	3.3	2.8	5.7
852520	Transmit-receive apparatus for radio, television, and so forth	0.3	15.8	15.6	44.3
852990	Parts for radio or television transmit-receive equipment, not elsewhere specified	7.2	6.0	−1.2	33.7
847330	Parts and accessories of data processing equipment, not elsewhere specified	1.9	5.1	3.2	48.2
441212	Plywood, 1 or 2 outer ply nonconifer, not elsewhere specified, ply <6 mm	0.6	45.4	44.8	21.8
850910	Domestic vacuum cleaners	0.0	41.4	41.4	47.2
854140	Photosensitive, photovoltaic, or light-emitting diode semiconductor devices	0.7	13.9	13.3	28.6

Source: Author's calculations using data from UN Comtrade.

10 products, Malaysia is not in a comfortable position in any products, because in no case does Malaysia have a commanding lead over China.

The situation with respect to Japan is broadly similar (see table 3.10). Again, almost all the products are electronic components, except for plywood products.[13] Only the plywood products are not under any threat from Chinese exports. The remaining eight products are either directly or partially threatened by China's exports.

The story in the EU15 is much the same, although the market shares of Malaysian products in the EU15 are quite small compared with those in the United States and Japan (see table 3.11). Apart from nonsurgical rubber gloves and sacks, most other exports from Malaysia to the EU are electronics and electronic components. In the EU15 market, all products, even nonsurgical rubber gloves, are under significant competitive pressure from China, except for a slight edge in monolithic integrated circuits.

Looking at Malaysia's top 10 exports to its three most important markets indicates that Malaysia is subject to competitive pressure from China virtually across the entire spectrum of key exports. Only in resource-based products (plywood products and rubber products) are Malaysian exporters relatively secure. Also, in comparative terms, Malaysia does not have a strong presence in the EU market for electronic components, its main export products.

---

[13] Most likely plywood is among the top 10 because of the value-to-weight (bulk) ratio.

Table 3.11 Dynamic Revealed Competitiveness Position for Malaysia's Top 10 Exports to the EU15, 2007

Product	Product name	Share of EU15's imports 1990	Share of EU15's imports 2007	DRCP Malaysia	DRCP China
847330	Parts and accessories of data processing equipment, not elsewhere specified	0.2	9.2	9.0	18.7
854211	Monolithic integrated circuits, digital	1.7	6.1	4.5	4.2
852520	Transmit-receive apparatus for radio, television, and so forth	0.7	2.0	1.3	24.8
847193	Computer data storage units	0.0	3.2	3.2	12.7
851790	Parts of line telephone or telegraph equipment, not elsewhere specified	0.0	4.3	4.3	24.8
852530	Television cameras	0.0	3.4	3.4	29.3
854140	Photosensitive, photovoltaic, or light-emitting diode semiconductor devices	8.3	3.0	−5.3	29.9
852990	Parts for radio or television transmit-receive equipment, not elsewhere specified	0.1	2.8	2.7	15.5
401519	Rubber gloves, other than surgical	35.2	43.7	8.5	9.1
392321	Sacks and bags (including cones) of polymers of ethylene	0.0	10.2	10.2	10.6

Source: Authors' calculations using data from UN Comtrade.

The analysis is static in the sense that it focuses on the top 10 exported products to three markets in 2007. To gauge the changing competitiveness of Malaysian exports to these markets, one must examine the shifts in market shares. Tables 3.12 to 3.14 list the fastest-growing exports from Malaysia (measured by the increase in market share in the three markets). In the U.S. market, a mix of products spanning many technology levels are gaining market share, such as glycerine, fatty acids, lumber, telecommunication equipment, and electronic components (see table 3.12). Many of these products have a commanding market share in the United States, although they represent a rather small fraction of Malaysia's exports, except for computers. Among them, five are partially threatened by China's exports, including computers, which represent 5 percent of Malaysia's exports.

The situation in Japan is quite different. Rather than electronics, the mix of Malaysian exports favors chemical- and resource-based products, such as palm oil, vegetable oils, and soaps (see table 3.13). These products have a large market share in Japan, and only one product is partially threatened by China's exports: vacuum cleaners. Malaysia's dynamic exports to the EU15 are similar—and are also mainly chemical- and resource-based products, along with electronic components (see table 3.14). In contrast to the Japanese market, Malaysian exports to the EU15 do not have dominant market shares, except for lumber. Those products that are partially threatened by China's exports are mainly electronic components. Malaysia still remains competitive in oscilloscopes.

Table 3.12 Malaysia's Most Dynamic Export Goods to the United States, 1990–2007

Product	Product name	Share of U.S. imports 1991	Share of U.S. imports 2007	DRCP China	DRCP Malaysia	Trade balance (US$ million)	Share of exports (%)	Technology classification
152090	Glycerol (glycerine), not elsewhere specified, including synthetic glycerol	1.0	52.9	n.a.	51.8	42.5	0.1	MT2
151919	Industrial monocarboxylic fatty acids, not elsewhere specified, acid oils	10.4	60.7	n.a.	50.3	n.a.	n.a.	RB1
903020	Cathode ray oscilloscopes, oscillographs	2.5	41.9	48.1	39.3	37.4	0.0	HT2
903040	Gain or distortion and cross-talk meters and so forth	0.1	35.6	n.a.	35.5	248.0	0.3	HT2
851939	Turntables, without record changers	1.4	33.2	31.6	31.8	n.a.	0.0	MT3
847120	Digital computers with central processing and input-output units	0.0	28.4	61.6	28.4	7,930.0	4.1	HT1
401692	Erasers (vulcanized rubber)	0.0	25.5	51.1	25.5	5.4	0.0	RB1
440721	Lumber, meranti red, meranti bakau, white lauan, and so forth	38.4	61.8	n.a.	23.4	9.1	0.2	RB1
902490	Parts and accessories of material testing equipment	0.0	20.7	8.3	20.6	6.6	0.0	HT2
152010	Glycerol (glycerine), crude and glycerol waters and lye	26.9	45.7	n.a.	18.8	1.6	0.0	MT2
730793	Butt weld fittings, iron or steel except stainless or cast	0.3	18.6	−17.9	18.3	25.4	0.0	MT2
903090	Parts and accessories, electrical measuring instruments	1.2	19.0	14.8	17.8	−133.0	0.2	HT2
291570	Palmitic acid, stearic acid, their salts and esters	41.5	58.7	0.1	17.2	38.1	0.1	MT2
851730	Telephonic or telegraphic switching apparatus	0.0	15.6	32.5	15.5	2,120.0	0.0	HT1
441211	Plywood 1 or 2 outer ply tropical hardwood, ply < 6 mm	11.5	26.7	32.0	15.2	n.a.	0.3	RB1

Source: Authors' calculations using data from UN Comtrade.
Note: n.a. = not applicable; HT1 = electronic and electrical products; HT2 = other high-technology products; MT2 = process industry; MT3 = engineering products; RB1 = agro-based products. Technology classification is based on Lall (2000).

Table 3.13 Malaysia's Most Dynamic Export Goods to Japan, 1990–2007

Product	Product name	Share of Japanese imports		DRCP		Trade balance (US$ million)	Share of exports (%)	Technology classification
		1990	2007	Malaysia	China			
151110	Palm oil, crude	0.4	99.9	99.5	n.a.	n.a.	0.7	RB1
180200	Cocoa shells, husks, skins, and waste	3.8	100.0	96.2	n.a.	−0.04	0.0	PP
151620	Vegetable fats, oils, or fractions hydrogenated, esterified	0.7	59.6	58.9	−1.7	14.30	0.8	RB1
903081	Electrical measurement recording instruments	0.3	56.7	56.4	n.a.	44.40	0.2	HT2
441211	Plywood 1 or 2 outer ply tropical hardwood, ply <6 mm	0.7	56.6	55.9	0.9	n.a.	0.3	RB1
854290	Parts of electronic integrated circuits and so forth	4.0	58.8	54.8	n.a.	−1,290.00	0.9	HT1
854190	Parts of semiconductor devices and similar devices	6.1	57.6	51.5	12.6	−59.00	0.3	HT1
852731	Radio-telephony receiver, with sound reproduce or record	11.3	56.8	45.6	28.2	153.00	0.2	MT3
720430	Waste or scrap tinned iron or steel	19.0	64.0	45.0	n.a.	n.a.	0.0	RB2
441212	Plywood, 1 or 2 outer ply nonconifer, not elsewhere specified, ply <6 mm	0.6	45.4	44.8	21.8	171.00	0.7	RB1
852711	Radio receivers, portable, with sound reproduce or record	12.8	57.6	44.8	17.7	56.30	0.1	MT3
340111	Soaps, for toilet use, solid	2.8	47.1	44.3	1.6	23.10	0.0	MT2
850910	Domestic vacuum cleaners	0.0	41.4	41.4	47.2	148.00	0.3	MT3
262030	Ash or residues containing mainly copper	0.1	40.9	40.8	n.a.	n.a.	0.0	RB2
180400	Cocoa butter, fat, oil	3.7	44.4	40.7	−3.2	n.a.	0.3	PP

Source: Authors' calculations using data from UN Comtrade.

Note: n.a. = not applicable; HT1 = electronic and electrical products; HT2 = other high-technology products; MT2 = process industry; MT3 = engineering products; PP = primary products; RB1 = agro-based products; RB2 = other resource-based products. Technology classification is based on Lall (2000).

Table 3.14 Malaysia's Most Dynamic Export Goods to the EU15, 1990–2007

Product	Product name	Share of EU15 imports (%) 1990	Share of EU15 imports (%) 2007	DRCP Malaysia	DRCP China	Trade balance (US$ million)	Share of exports (%)	Technology classification
440721	Lumber, meranti red, meranti bakau, white lauan, and so forth	39.7	78.8	39.1	0.1	303.0	0.2	RB1
230660	Palm nut or kernel oil cake and other solid residues	15.6	46.2	30.5	n.a.	n.a.	0.1	PP
847021	Electronic calculators, printing, external power	1.2	27.2	26.0	20.9	14.0	0.0	HT1
151110	Palm oil, crude	16.1	38.5	22.4	n.a.	n.a.	0.7	RB1
852739	Radio-broadcast receivers, not elsewhere specified	4.1	26.2	22.2	29.3	121.0	0.3	MT3
151911	Stearic acid	3.8	25.7	21.9	n.a.	48.7	0.2	RB1
401511	Rubber surgical gloves	24.5	44.8	20.2	−6.0	238.0	0.1	LT1
151919	Industrial monocarboxylic fatty acids, not elsewhere specified, acid oils	1.4	18.8	17.4	n.a.	n.a.	n.a.	RB1
400129	Natural rubber in other forms	14.0	29.8	15.8	0.0	n.a.	0.0	PP
903020	Cathode ray oscilloscopes, oscillographs	0.7	13.3	12.6	3.4	23.2	0.0	HT2
392321	Sacks and bags (including cones) of polymers of ethylene	0.0	10.2	10.2	10.6	303.0	0.2	LT2
847330	Parts and accessories of data processing equipment, not elsewhere specified	0.2	9.2	9.0	18.7	3,720.0	5.9	HT1
401519	Rubber gloves, other than surgical	35.2	43.7	8.5	9.1	311.0	0.8	LT1
140120	Rattan used primarily for braiding	3.3	11.3	8.1	44.0	573.6	0.0	PP
291570	Palmitic acid, stearic acid, their salts and esters	0.3	8.0	7.7	18.3	18.3	0.1	MT2

Source: Authors' calculations using data from UN Comtrade.
Note: n.a. = not applicable; HT1 = electronic and electrical products; HT2 = other high-technology products; LT1 = textiles, garments, and footwear; LT2 = other low-technology products; MT2 = process industry; MT3 = engineering products; RB1 = agro-based products; PP = primary products. Technology classification is based on Lall (2000).

What are the products in which Malaysia is losing market share the most? Surprisingly, in the U.S. market, crude palm oil is the one that has lost the most market share between 1990 and 2006. Very likely this loss is because of the rise of Indonesia as an exporter of palm oil. However, crude palm oil gained the most market share in the Japanese and EU15 markets, offsetting the lost market share in the United States. Many products that lost market share in the United States are resource-based and low-tech items, such as textile products and low-end electronics. In some of these products (including some resource-based ones), Chinese products are displacing Malaysian products in the United States. Similarly, in the Japanese market, Malaysia's resource-based and low-tech products lost market shares. However, unlike the case in the U.S. market, this loss was not solely because of the rise of China. China gained market share in only four products, and China does not have any presence in the other products in the Japanese market. In the EU15 market, resource-based products represent the majority of products that lost market share, in addition to low-tech items such as textiles. Again, the presence of China is not significant in the EU15 for those products in which Malaysia lost market shares.

The movement in Malaysia's dynamic revealed competitiveness position (DRCP) by technology level indicates that its exports are losing ground (see tables 3.15 to 3.17). The DRCP is negative in all technology classes except for other resource-based products (RB2) and other high-tech products (HT2). Meanwhile, China substantially improved its DRCP in all technology levels (see table 3.15), although less so in primary products (PP), other high-tech products (HT2), and

Table 3.15 **Dynamic Revealed Competitiveness Position by Technology Level in U.S. Market: Malaysia versus China, 1991–2007**

Technology classification	Malaysia			China		
	1991–96	1997–2000	2001–07	1991–96	1997–2000	2001–07
HT1	3.34	0.73	−0.51	3.54	3.04	23.75
HT2	0.36	−0.16	0.30	2.29	0.16	0.55
LT1	0.23	−0.27	−0.28	4.80	−1.11	18.31
LT2	0.61	−0.37	−0.09	9.46	4.58	12.97
MT1	0.01	0.00	0.00	0.23	0.32	1.76
MT2	0.52	0.00	−0.18	0.32	2.45	6.62
MT3	0.67	−0.24	−0.27	3.46	2.97	8.28
PP	−0.17	0.04	−0.03	−0.36	−0.21	0.26
RB1	0.03	0.01	0.71	0.83	1.04	6.66
RB2	0.19	0.00	−0.19	−17.31	8.74	36.90

Source: Authors' calculations using data from UN Comtrade.
Note: HT1 = electronic and electrical products; HT2 = other high-technology products; LT1 = textiles, garments, and footwear; LT2 = other low-technology products; MT1 = automotive products; MT2 = process industry; MT3 = engineering products; PP = primary products; RB1 = agro-based products; RB2 = other resource-based products. Technology classification is based on Lall (2000).

Table 3.16 Dynamic Revealed Competitiveness Position by Technology Level in Japanese Market: Malaysia versus China, 1991–2007

Technology classification	Malaysia			China		
	1990–96	1997–2000	2001–07	1990–96	1997–2000	2001–07
HT1	4.03	2.35	−3.55	6.32	1.79	19.92
HT2	0.78	0.46	−0.22	2.64	1.35	3.93
LT1	0.23	−0.29	−0.05	28.92	11.76	6.43
LT2	1.66	0.42	−0.63	12.87	5.52	16.49
MT1	0.11	0.02	0.14	1.49	2.38	8.13
MT2	0.96	0.99	−0.26	1.99	−0.47	8.85
MT3	4.00	−0.19	−1.85	9.37	4.89	15.60
PP	0.10	0.14	−0.13	0.21	−0.60	−2.39
RB1	−2.26	−1.66	1.23	6.16	3.17	5.26
RB2	0.40	−0.30	−0.12	−9.45	8.84	−1.66

Source: Authors' calculations using data from UN Comtrade.
Note: HT1 = electronic and electrical products; HT2 = other high-technology products; LT1 = textiles, garments, and footwear; LT2 = other low-technology products; MT1 = automotive products; MT2 = process industry; MT3 = engineering products; PP = primary products; RB1 = agro-based products; RB2 = other resource-based products. Technology classification is based on Lall (2000).

Table 3.17 Dynamic Revealed Competitiveness Position by Technology Level in EU15's Market: Malaysia versus China, 1991–2007

Technology classification	Malaysia			China		
	1990–96	1997–2000	2001–07	1990–96	1997–2000	2001–07
HT1	2.37	−0.27	0.14	1.71	2.27	13.12
HT2	0.14	−0.01	−0.05	0.76	0.04	−0.05
LT1	0.27	−0.06	−0.12	2.87	3.08	11.54
LT2	0.17	0.13	−0.01	2.37	2.70	5.79
MT1	0.09	−0.05	0.01	0.04	0.10	0.48
MT2	0.18	0.07	−0.07	0.32	0.56	1.59
MT3	0.52	−0.05	−0.12	1.54	1.46	4.23
PP	0.05	−0.04	0.03	0.16	0.05	0.07
RB1	0.10	0.00	0.12	0.29	0.30	1.32
RB2	0.10	−0.03	−0.02	−10.42	9.36	13.90

Source: Authors' calculations using data from UN Comtrade.
Note: HT1 = electronic and electrical products; HT2 = other high-technology products; LT1 = textiles, garments, and footwear; LT2 = other low-technology products; MT1 = automotive products; MT2 = process industry; MT3 = engineering products; PP = primary products; RB1 = agro-based products; RB2 = other resource-based products. Technology classification is based on Lall (2000).

automotive products (MT1). In the Japanese market, Malaysia lost market shares in many technology categories, except for automotive products (MT1) and agro-based products (RB1) (see table 3.16). Changes in the EU15 market are small (see table 3.17). Again, China is increasing its market shares in all technology levels, except for those in primary products and resource-based products in the Japanese market and other high-tech products (HT2) in the EU15. Clearly, China is expanding the fastest in high-tech products, followed by low-tech products. In the Japanese market, China is also rapidly expanding its share in medium-tech products.

Export Diversification or Specialization

One strand of research maintains that export diversification and an increase in the unit value of exports are the principal avenues for a country trying to raise export earnings.[14] Empirical research on the composition of exports (with an increasing emphasis on China) has brought to light a number of interesting patterns. First, lower-income countries can stimulate export growth by diversifying their exports and exporting to markets where demand is growing more rapidly. Low- and medium-tech exports of primary and processed commodities are the rule at this stage of development. With rising incomes, export growth through diversification is less marked but may remain significant, to be followed by greater specialization in specific export categories (Hummels and Klenow 2005). Middle- and high-income countries in East Asia are either specializing or beginning to specialize; even China is specializing (Amiti and Freund 2007). A sizable percentage of exports of middle-income East Asian economies (electronic products, for example) falls in the high-tech categories and overlaps with exports from the United States or other industrial countries (Schott 2001). The only difference is that, as with China, the products are likely to be of lower technological sophistication and quality that are reflected in the unit costs (Schott 2006; Xu 2007). Presumably, if the middle-income countries continue to specialize, the way forward will be to raise the quality of their products and move toward the upper end of the product spectrum. This process could intensify competition and erode profitability unless producers also diversify into ever-narrower subspecializations. Absent a new technological epoch comparable to the information technology and electronics revolution, the trend toward specialization will subject all producers to greater pressure, which will only be eased by faster technological change in current areas of export concentration.

The emergence of China as the world's second-largest exporter in 2007 and a major importer is beginning to affect the opportunities available to other countries, with notable consequences (Dimaranan, Ianchovichina, and Martin 2007). Increasingly, China is beginning to dominate the export markets for finished consumer and electronic products, including, for example, a wide range of consumer electronics. Lower labor costs aided by scale economies have given

---

[14]What policies or market forces can trigger and promote diversification remain uncertain, and these lacunae hamper policy making (Klinger and Lederman 2006).

Figure 3.10 Imports of Electronic Components by China from East Asian Countries, 1995–2007

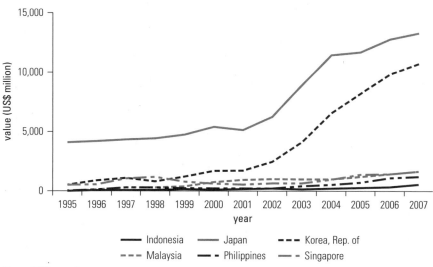

Source: UN Comtrade.

Chinese firms a comparative advantage in product assembly. This advantage has increased China's imports of parts and components from other East Asian economies and is the basis for a mutually beneficial division of labor between China and its neighbors thus far. However, this relationship is evolving as Chinese producers and MNCs producing in China integrate backward into the production of parts and components, starting with the simpler ones, and MNCs increasingly try to source parts for their worldwide operations from fewer and larger suppliers. This trend raises the barriers to new entrants. It will likely be exacerbated by the global recession of 2008 and 2009, which will affect the weaker producers apart from encouraging greater clustering of suppliers and assemblers. This change could be detrimental for countries such as Indonesia, Malaysia, the Philippines, and Thailand over the medium run if major component suppliers and subsidiaries of MNCs relocate to China and local firms are hampered from entering by their smallness and the time-consuming certification procedures enforced by MNCs. Although China continues to import a large volume of intermediate inputs, more of them are now complex and higher-technology items (see Cui and Syed 2007),[15] which tend to be sourced from Japan, Korea, and Thailand, as can be seen from figure 3.10. This is also the

---

[15] Gaulier, Lemoine, and Ünal-Kesenci (2007) did not find evidence that Chinese exports were displacing those from Malaysia, the Philippines, and Thailand, and at least through 2003, the Southeast Asian countries were gaining export shares in the upper-tier product categories.

Figure 3.11 Imports of Electronic Components by China from East Asian Countries, Excluding Japan and Korea, 1995–2007

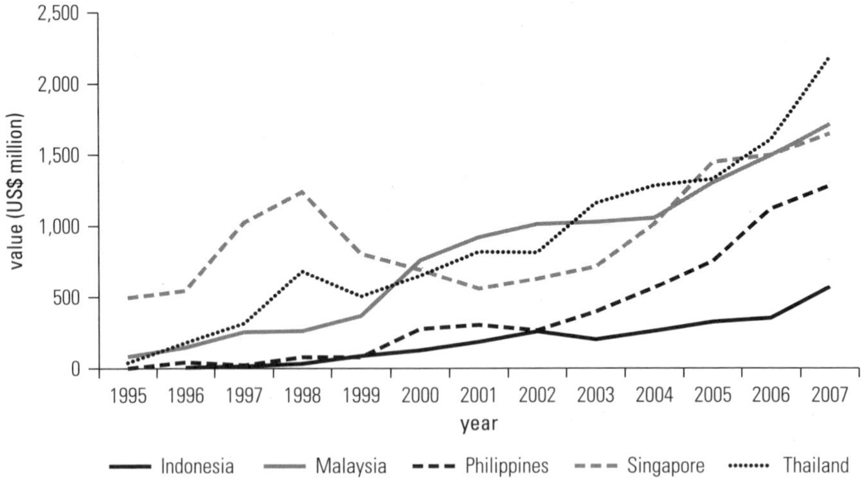

Source: UN Comtrade.

reason the exports from these economies fell so sharply starting in the final quarter of 2008. The other Southeast Asian countries are experiencing a slower pace of increase in their exports of electronic products to China and could, in fact, face stiffer competition from Chinese producers in third-country markets unless they upgrade their exports. From figures 3.10 and 3.11, one can see that Malaysia's exports of electronics and engineering products to China are expanding, but less so than those of Japan, Korea, and Thailand, and that the acceleration of lower-tech commodity components exports most favors the Philippines. Figure 3.12 indicates that the share of China's imports of electronics from Malaysia first expanded between 1995 and 2000 and then contracted between 2000 and 2007. Meanwhile, the share of China's exports of electronic components to Malaysia has grown since 1995 (see figure 3.13). This finding is consistent with trends in investment within the electronics industry (see table 3.18). During 1996 to 2000, the components subsector dominated the investment in electronics, accounting for three-quarters of investment. By 2001 to 2005, the share of investment in components had declined to 55 percent. Meanwhile, investment in industrial electronics tripled and came to account for one-third of investment in electronics (Yusof and Bhattasali 2008).

New Exports

How do Malaysia's efforts at diversifying exports compare with those of other countries? Is Malaysia exporting higher-value and technologically sophisticated products? And are these exports mainly by MNCs or by Malaysian firms also?

Analyzing Comparative Advantage and Industrial Change   63

Figure 3.12   Imports of Electronic Components by China, According to Economy of Origin, 1995–2007

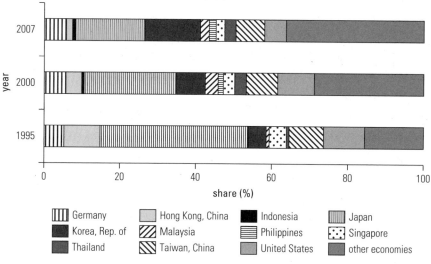

Source: UN Comtrade.

Figure 3.13   Exports of Electronic Components by China According to Economy of Destination, 1995–2007

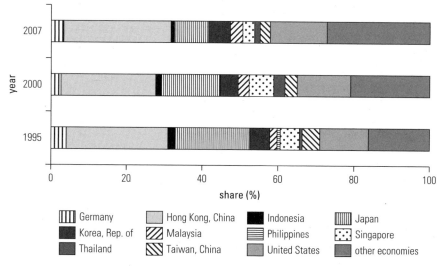

Source: UN Comtrade.

Table 3.18 Investments in Electrical and Electronics Industry, 1996–2005

Product category	1996–2000 Value (RM billion)	1996–2000 Share (%)	2001–05 Value (RM billion)	2001–05 Share (%)	1996–2005 Value (RM billion)	1996–2005 Share (%)
Total	41.0	100.0	43.3	100.0	84.3	100.0
Electronics products	36.5	89.0	40.7	93.9	77.2	91.6
Components	30.5	74.4	24.0	55.3	54.5	64.7
Industrial	4.6	11.2	14.0	32.3	18.6	22.0
Consumer	1.4	3.4	2.7	6.3	4.1	4.9
Electrical products	4.5	11.0	2.6	6.1	7.1	8.4

Source: Yusof and Bhattasali 2008.

The numbers of new commodities entering the export baskets of selected Asian countries during 1996 to 2007 are shown in table 3.19. Indonesia led the field with 299 commodities. Malaysia occupied the middle ground with 137. Singapore had the fewest—just 17. Some commodities cease to be exported, and a number of commodities enter and exit the export mix in a cyclical fashion. Table 3.20 provides a clearer picture of the overall situation. Surprisingly, the composition of China's exports varied the least, and just 8.2 percent of exported products were subject to cyclical ins and outs. China, along with the more advanced economies

Table 3.19 Number of Commodities That Entered or Exited Selected Asian Export Baskets, by Category

Economy	Steady	New	Disappeared	Appeared and vanished	Cyclical	Period
China	4,142	87	301	22	407	1995–2007
Indonesia	2,387	299	158	90	2,005	1995–2007
Japan	3,978	44	315	48	538	1995–2007
Korea, Rep. of	3,495	68	260	77	1,013	1995–2007
Malaysia	3,244	137	129	101	1,325	1995–2007
Philippines	1,388	232	181	466	1,909	1996–2007
Singapore	3,947	17	396	27	590	1995–2007
Taiwan, China	3,887	60	88	69	718	1997–2007
Thailand	3,222	154	211	101	1,262	1995–2007

Source: UN Comtrade.
Note: "Steady" means that a commodity has positive export values for all years; "new" means that a country did not export a commodity at the beginning of the period, but it started to export in later years; "disappeared" means a country stopped exporting a commodity; "appeared and vanished" means a country started to export a commodity sometime after 1995 but stopped exporting before the end of the period; and "cyclical" means that a commodity was exported irregularly during the period.

Table 3.20 Percentage of Commodities That Entered or Exited Selected Asian Export Baskets, by Category

Economy	Steady (%)	New (%)	Disappeared (%)	Appeared and vanished (%)	Cyclical (%)	Period
China	83.5	1.8	6.1	0.4	8.2	1995–2007
Indonesia	48.3	6.1	3.2	1.8	40.6	1995–2007
Japan	80.8	0.9	6.4	1.0	10.9	1995–2007
Korea, Rep. of	71.1	1.4	5.3	1.6	20.6	1995–2007
Malaysia	65.7	2.8	2.6	2.0	26.8	1995–2007
Philippines	33.2	5.6	4.3	11.2	45.7	1996–2007
Singapore	79.3	0.3	8.0	0.5	11.9	1995–2007
Taiwan, China	80.6	1.2	1.8	1.4	14.9	1997–2007
Thailand	65.1	3.1	4.3	2.0	25.5	1995–2007

Source: UN Comtrade.
Note: As a proportion of the total number of exported commodities by the country, over the period indicated. "Steady" means that a commodity has positive export values for all years; "new" means that a country did not export a commodity at the beginning of the period, but it started to export in later years; "disappeared" means a country stopped exporting a commodity; "appeared and vanished" means a country started to export a commodity sometime after 1995 but stopped exporting before the end of the period; and "cyclical" means that a commodity was exported irregularly during the period.

of Japan; Korea; Singapore; and Taiwan, China, had fewer new entries than exits and relatively low percentages of cyclical exports.[16]

As noted by Imbs and Wacziarg (2003), it is the high-income countries that are becoming more specialized. At the other extreme are lower-middle-income countries, such as India and Indonesia, that are diversifying rapidly. Malaysia appears to be on the cusp. New commodities just outnumber products disappearing from the export basket. However, the cyclical component is fairly large, suggesting that the export sector is still quite fluid and at an intermediate stage of specialization. The surprise is China. Its export dynamics are beginning to resemble those of a high-income country.

What are some of the fastest-growing "new" exports (as distinct from the fastest-increasing exports reported in table 3.3) from Malaysia? Table 3.21 lists the 20 items that registered the highest growth in value. They are a mixed bag: mostly chemicals and processed products. Whether they are the outcome of process

---

[16] The growth of exports from a country comes mainly from intensive margin (increase in exports to the same markets) rather than from expansion of the export market. Developing countries seem to have difficulty sustaining a trade flow with a country (high exit rate), although the entry rate is high. This finding suggests that firms—especially small ones—in developing countries export small quantities first to test the export market and subsequently decide to exit. These firms may need better market information if they are to survive in the export market (Brenton, Newfarmer, and Walkenhorst 2008).

Table 3.21 Fastest-Growing "New" Exports from Malaysia, 2000–07

Product	Product name	Trade value, 2007 (US$ million)	Annual growth rate, 2000–07 (%)	Year entered
722692	Cold-rolled alloy-steel, not elsewhere specified, not further worded, <600 mm wide	2.24	146.4	1996
290241	O-xylene	1.88	146.2	1998
290941	2,2′-oxydiethanol(diethylene glycol)	4.53	145.3	1997
960891	Pen nibs, nib points, not elsewhere specified	6.28	131.2	1996
741410	Endless bands of copper wire for machinery	1.16	127.6	1997
271600	Electrical energy	120.96	87.9	2000
370710	Sensitizing emulsions	0.32	68.5	1996
284990	Carbides, except calcium and silicon	0.88	66.0	1996
400239	Halo-isobutene-isoprene rubber (CIIR/BIIR)	0.09	66.0	1997
650300	Felt hats and other felt headgear	0.28	65.4	1996
270119	Coal, except anthracite or bituminous, not agglomerated	6.57	62.8	1996
293339	Heterocyclic compounds with unfused pyridine ring, not elsewhere specified	0.53	62.2	1996
711220	Waste or scrap containing platinum as the sole precious metal	12.50	57.7	1997
291241	Vanillin (4-hydroxy-3-methoxybenzaldehyde)	0.06	57.2	1998
290260	Ethylbenzene	4.26	55.1	1998
284020	Borates of metals, except refined borax	4.71	52.0	1996
580123	Woven weft pile cotton fabric, not elsewhere specified, width > 30 cm	10.17	46.5	1997
282619	Fluorides of metals, except ammonium, sodium, aluminum	0.07	46.2	1996
481131	Paper, > 150 g/m^2, bleached, plastic-coated or plastic-impregnated	4.33	42.6	1996
230220	Rice bran, sharps, other residues	0.18	40.3	1997

Source: UN Comtrade.

innovations that have raised the productivity of exporting firms is difficult to tell. However, they are of the low-tech variety and mirror the feedback from firms on the declining skill and technology intensity of production.

The fastest-growing electronics exports from Malaysia with regard to value are listed in table 3.22. Virtually all the products are being produced by MNCs rather than by Malaysian-owned firms. Most can be classified as high-tech commodities rather than as innovative products incorporating advances in knowledge. Malaysian firms specialize in precision machining, in manufacturing the simpler components, in making plastic and metal parts, and in making automation equipment. By comparison, Thailand's fastest-growing electronics exports cover many products for

Table 3.22 Top 10 Fastest-Growing Electronics and Electrical Exports from Malaysia, 2000–07

Product	Product name	Trade value, 2000 (US$)	Trade value, 2007 (US$)	Annual growth rate, 2000–07 (%)
900921	Photocopying equipment with an optical system, not elsewhere specified	1,400	141,000,000	519.3
850530	Electromagnetic lifting heads	22,700	10,000,000	238.7
902229	Nonmedical apparatus using alpha, beta, or gamma radiation	121,600	21,600,000	209.7
854320	Signal generators	2,915,700	230,000,000	186.6
846921	Typewriters, electric, >12 kg, nonautomatic	109,800	8,146,500	185.0
841239	Pneumatic power engines or motors, except linear acting	208,900	11,700,000	177.8
853120	Indicator panels incorporating electronic displays	5,068,900	211,000,000	170.3
850163	Air-conditioning generators, output 375–750 kVA	21,000	852,100	169.7
850422	Liquid dielectric transformers, 650–10,000 kVA	183,700	5,256,200	161.5
902230	X-ray tubes	122,500	3,397,100	160.7

Source: UN Comtrade.

the hard disk drive industry (see table 3.23). The Philippines' fastest-growing electronic exports include semiconductor products and batteries (see table 3.24). China's 10 fastest-growing exports in electronics and electrical products include a number of key components for communications equipment and other machinery (see table 3.25). None of these fastest-growing commodities shows much overlap among countries, except for radio navigational aid apparatus between China and Thailand.

The rapidly growing exports from China, Korea, and Malaysia reflect the strength of current market demand, but whether they can help identify the future leading sectors for Malaysian firms is open to conjecture because they spring from technologies introduced by MNCs and do not represent indigenous technological capacity.[17] With rising incomes, not only is there a trend toward

---

[17] For example, the growth of manufactured exports, especially of electronic products, may not have been the result of improved technological capabilities nor of the accumulation of human capital. Rather, it appears simply to reflect external changes, especially the sharp appreciation of the Japanese yen after the Plaza Accord in 1985, which prompted large inflows of foreign direct investment into Malaysia (and Thailand) to take advantage of lower costs. By 1996, heavy investment in real estate resulted in an overheating economy with rising wage costs and transport bottlenecks in Malaysia and Thailand. In turn, this development resulted in a rising real effective exchange rate, a loss of competitiveness, and slower export growth, which contributed to the Asian crisis (Corden 2007; Reinhardt 2000).

Table 3.23 Top 10 Fastest-Growing Electronics and Electrical Exports from Thailand, 2000–07

Product	Product name	Trade value, 2000 (US$)	Trade value, 2007 (US$)	Annual growth rate, 2000–07 (%)
840110	Nuclear reactors	3,800	31,500,000	363.3
847193	Computer data storage units	1,508,900	9,490,000,000	348.8
852691	Radio navigational aid apparatus	32,200	82,400,000	306.8
853080	Electric signal, safety and traffic controls, not elsewhere specified	18,800	45,900,000	304.8
847310	Parts and accessories of typewriters, word processors	37,300	46,100,000	276.6
850230	Electric generating sets, not elsewhere specified	1,036,800	407,000,000	234.7
902229	Nonmedical apparatus using alpha, beta, or gamma radiation	700	218,300	227.4
854012	Monochrome cathode ray picture tubes, monitors	400	99,900	220.2
841239	Pneumatic power engines or motors, except linear acting	28,700	5,116,700	209.7
854310	Particle accelerators	28,300	4,007,500	202.9

Source: UN Comtrade.

Table 3.24 Top 10 Fastest-Growing Electronics and Electrical Exports from the Philippines, 2000–07

Product	Product name	Trade value, 2000 (US$)	Trade value, 2007 (US$)	Annual growth rate, 2000–07 (%)
854590	Battery carbons and carbon electrical items, not elsewhere specified	600	541,200	266.8
847193	Computer data storage units	3,234,200	2,080,000,000	251.9
854129	Transistors, except photosensitive, >1 watt	180,700	106,000,000	248.6
853090	Electric signal, safety and traffic controller parts	11,500	2,080,800	210.1
854390	Parts of electrical machines and apparatus, not elsewhere specified	10,600	1,635,400	205.4
853190	Parts of electric sound and visual signaling apparatus	89,500	13,600,000	204.9
902290	Parts and accessories for radiation apparatus	81,300	10,000,000	199.0
854219	Monolithic integrated circuits, except digital	632,800	57,600,000	190.5
851290	Parts of cycle and vehicle light, signal, and similar equipment	69,900	4,678,700	182.3
850133	DC motors, DC generators, output 75–375 kW	100	3,000	179.5

Source: UN Comtrade.

Table 3.25 Top 10 Fastest-Growing Electronics and Electrical Exports from China, 2000–07

Product	Product name	Trade value, 2000 (US$)	Trade value, 2007 (US$)	Annual growth rate, 2000–07 (%)
852691	Radio navigational aid apparatus	879,600	2,060,000,000	303.0
854042	Klystron tubes	800	836,200	269.8
852530	Television cameras	25,800,000	10,600,000,000	236.4
902221	Medical apparatus using alpha, beta, or gamma radiation	9,000	2,679,800	225.8
847230	Machinery for processing mail of all kinds	20,900	6,173,500	225.3
850164	Air-conditioning generators, output > 750 kVA	405,600	100,000,000	219.7
851730	Telephonic or telegraphic switching apparatus	101,000,000	15,200,000,000	204.7
902219	Nonmedical x-ray equipment	2,011,000	262,000,000	200.5
850220	Generating sets, with spark ignition engines	11,500,000	919,000,000	187.1
847120	Digital computers with central processing and input-output units	906,000,000	66,700,000,000	184.8

Source: UN Comtrade.

greater specialization in exports, but also countries tend to move toward the high-priced ends of their existing product categories by improving quality, design, and technology.

Quality and Technological Level of Tradable Products

Unit values of commodities—that is, price per unit of output of a commodity—are used to assess the "quality" of a product. The underlying assumption is that higher-priced items are superior, in their technological content or sophistication. However, meaningful comparisons of unit values can be made only at a disaggregated level—that is, by comparing unit values of a commodity across countries, or within the same country but over time. This limitation arises primarily from differences in the metrics used to measure the quantity of each commodity's output. For this study, annual data on the value and quantity of export of commodities at the six-digit harmonized system (HS) 1988–92 level were extracted from the United Nations Commodity Trade Statistics Database (UN Comtrade).[18]

---

[18] Further refinement of the product description at the 10-digit level is available from the HS 2002 classification. The constraint to using such a level of disaggregation is that time-series data on exports and imports are available only from 2002. The data used here cover the period from 1995 to 2006.

Given the heterogeneity of exported products and the range of metrics used to measure export volumes, deriving a unit value index is extremely problematic (Kaplinsky and Paulino 2005). Moreover, even when apparently homogeneous commodities are compared, the data are "noisier" from some countries than from others.

Unit values were generated by dividing the export values by the quantity for each product exported in a selected year and for each country. At the six-digit commodity level, the data were subject to substantial variation with regard to the unit in which the quantity of output was measured, both across countries and within a country, over time. About 35 percent of the commodities in the electronics and electrical category had discrepancies in the units in which a product was measured across the eight countries sampled. Even after restricting the comparison to the products that were recorded to have been measured in the same units, across countries, variability in the unit values remained, both within a commodity category and across countries (and years), and within a country (across years) for a commodity. The median value of the range of these unit values within a commodity was around US$530 and was US$70 even within a country for the same commodity.

Figures 3.14 through 3.19 display the trends in unit values for a number of widely traded electronic components: digital monolithic integrated circuits, monolithic integrated circuits (except digital), hybrid integrated circuits, electronic integrated circuits, parts of integrated circuits, and cathode ray tubes and valves. These products were chosen because their production technology is fairly standardized; hence, one can expect that their unit values across countries will be similar. But the figures

Figure 3.14 Unit Values of Digital Monolithic Integrated Circuits, 1995–2007

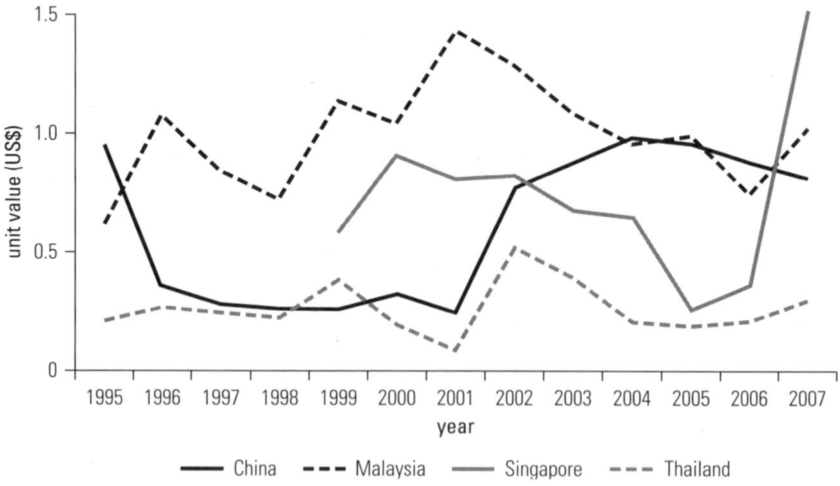

Source: UN Comtrade.

Figure 3.15 Unit Values of Nondigital Monolithic Integrated Circuits, 1995–2007

Source: UN Comtrade.

Figure 3.16 Unit Values of Hybrid Integrated Circuits, 1995–2007

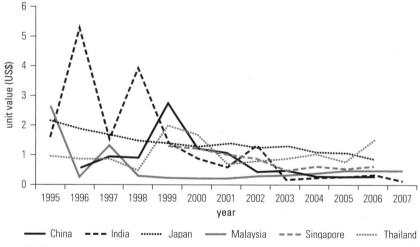

Source: UN Comtrade.

show that there is even variation in countries that are outliers with regard to having unit values that differed significantly from the remaining countries in the sample. Indonesia and Korea had higher unit values for four out of the five integrated circuit commodities considered, with Thailand (as well as Japan, to some extent) being the outlier in the fifth product considered. As for color cathode ray television

Figure 3.17 Unit Values of Electronic Integrated Circuits, 1995–2007

Source: UN Comtrade.

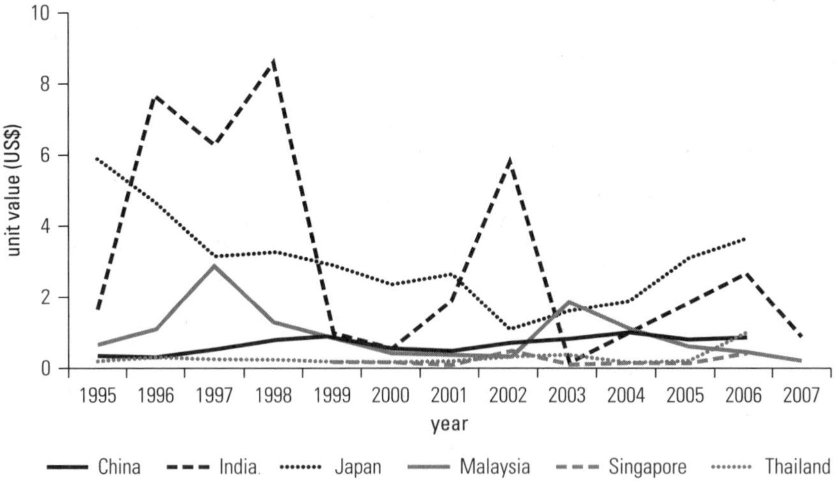

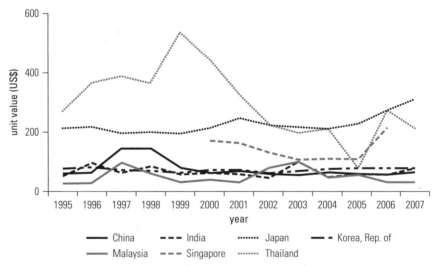

Figure 3.18 Unit Values of Parts of Electronic Integrated Circuits and Similar Items, 1995–2007

Source: UN Comtrade.

pictures and monitors, the unit values prevailing in Indonesia, Japan, and Korea were significantly smaller than in the other countries sampled. Hence, data for unit values of Indonesian and Korean exports were dropped for some of these products. In one instance, Thai data also had to be excluded for the same reason.

Figure 3.19 Unit Values of Color Cathode Ray Television Picture Tubes and Similar Items, 1995–2007

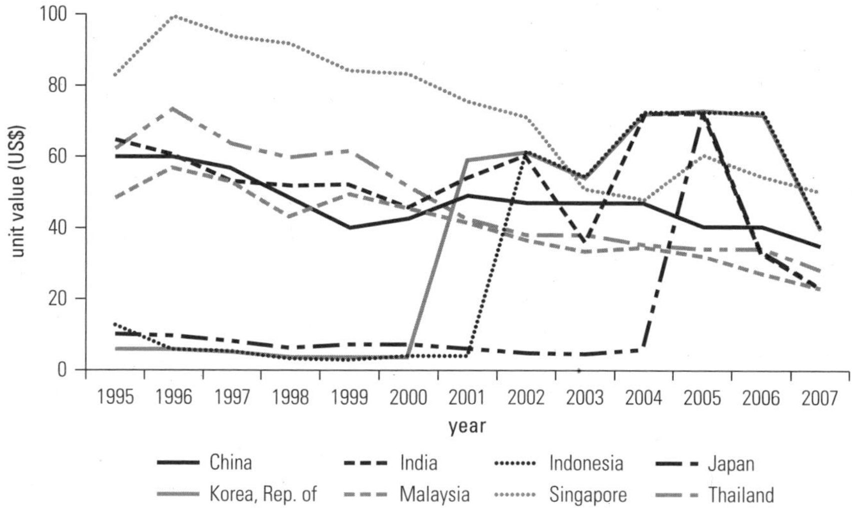

[Figure: Line chart showing unit values (US$) from 1995 to 2007 for China, India, Indonesia, Japan, Korea Rep. of, Malaysia, Singapore, and Thailand. Y-axis ranges from 0 to 100 US$.]

Source: UN Comtrade.

Keeping in mind the concerns regarding the "quality" of the data that have been extracted from UN Comtrade, one can infer (see figure 3.14) that unit values for digital monolithic integrated circuits improved somewhat for Malaysia from 1995 to 2007, in contrast to the levels prevailing in China and Thailand. Similarly, figure 3.15 shows that unit values (even though declining after 2000) for nondigital monolithic integrated circuits from Malaysia were higher than in most of the other countries (except India), over the entire period from 1995 to 2007. As for hybrid integrated circuits, unit values from Malaysia were comparable to those of most other East Asian countries in the sample, as seen in figure 3.16. Figure 3.17 shows that although the unit values for electronic integrated circuits were declining in Malaysia, they were comparable to those of most other Asian economies in the sample. From figure 3.18, it is apparent that for parts of electronic integrated circuits, unit values in Malaysia were much lower than those in Japan, but were comparable to other Asian countries in the sample, albeit with significant fluctuation. Finally, for color cathode ray television tubes, figure 3.19 indicates that unit values in Malaysia were on a slightly declining trend like some of the other countries in the sample. A few countries (Indonesia and Korea) show an improvement from 2003 to 2005; however, the reasons are difficult to pin down.

Overall, at least for this small set of products, one can conclude (with all the due qualifications) that Malaysian firms were technologically competitive with firms in other East Asian economies at similar stages of development. One can also conclude that, although they oscillate a fair amount, the trends in unit value are either flat, as with monolithic integrated circuits (except digital), or downward, as with hybrid integrated circuits. In many cases, Malaysia's unit value for a product is decreasing, except for digital monolithic integrated circuits. Consequently, from this sample of electronic products, there is no consistent evidence that the technological sophistication of Malaysia's exports or their quality has improved sufficiently to raise its unit export values over the period from 1995 to 2007.[19]

Because the calculated unit values exhibited considerable variability in all possible dimensions, another measure of technological sophistication was constructed following Hausmann, Hwang, and Rodrik (2007). These measurements map to the current HS 1988–92 product classification. For each product, a weighted average of the GDP per capita of countries exporting that product is calculated to assign a value (PRODY) that is a proxy for quality. The weights denote the revealed comparative advantage of each country exporting the product. Chandra, Boccardo, and Osorio (2007) calculated this measure of the productivity level (based on the countries that dominate exports) associated with a product at the Standard Industrial Trade Classification revision 2 (SITC2) four-digit level for approximately 1,006 products. Within the electronic and electrical group of commodities, the HS 1988–92 six-digit level was matched with the SITC2 four-digit level using a concordance table available in UN Comtrade. A total of 205 products within the electronic and electrical category at the six-digit level could be mapped to their corresponding product values. From 1995 to 2006, on average, Taiwan, China, exported electronics and electrical commodities with a product value of US$14,515, the highest for the group. Indonesia, Korea, and Malaysia followed with values of US$14,511, US$14,485, and US$14,482, respectively. Japan, Singapore, and Thailand were next with an average productivity value of approximately US$14,482. Finally, China's exports averaged a productivity value of US$14,479.

The basket of electronic and electrical commodities exported by the countries sampled was further differentiated by their frequency of appearance and sustained presence on the export market from 1995 to 2006. The implicit assumption behind this categorization is that more profitable commodities stay on the export market longer than those with lesser value added. Three specific subcategories are of particular interest for assessing the relative technological competitiveness of Malaysia's exports. At the HS 1988–92 six-digit level, products are identified as "steady" if a positive export value was recorded for all 12 years. A product was categorized as "new" if the beginning year of a positive export value of the

---

[19] South Asian countries are in a similar situation, with unit values of exports increasing only minimally during 1991 to 2002. Much of the growth in exports was the result of a decline in relative unit costs rather than because of an increase in quality (Brunner and Cali 2006).

commodity was later than 1995, but the commodity stayed on the export market continuously through 2006. The third category of relevance was of products identified as "cyclical," meaning that their frequency of "appearance" (with a positive export value) was irregular.

The sum of the productivity values of the steady commodities in this sector made it possible to gauge the technological achievements of the exporting country before 1995. This statistic captures the level of technological sophistication of the products that were "discovered" previously and that have been exported since. The arithmetic sum rather than the mean of the productivity values is examined, because the number of commodities in each of these categories matters, not just their productivity values. Consider the following hypothetical example: Malaysia has three electronics and electrical commodities in the steady category for the period from 1995 to 2006. Indonesia, on the other hand, has two electronics and electrical commodities in the steady category for the same period. The corresponding productivity value for each of these steady commodities is 100. Hence, the sum of the productivity values of the steady commodities for Malaysia is 300, whereas it is 200 for Indonesia. To put it differently, all other things being equal, if Malaysia exports a larger number of electronics and electrical products in the steady category than Indonesia, one may infer that Malaysia has greater technological capacity than Indonesia.

Japan and Singapore ranked highest for this set of steady electronic and electrical items, at US$2.84 million and US$2.73 million, respectively. China, Korea, and Malaysia were next at US$2.66 million, US$2.58 million, and US$2.57 million, respectively. Taiwan, China; India; and Thailand followed, with total productivity values of US$2.56 million, US$2.55 million, and US$2.54 million, respectively. Last, exports of stable electronic and electrical items by Indonesia had the lowest productivity levels of US$1.74 million. These numbers imply that Malaysia has made reasonable progress toward the technological frontier for electronic and electrical products that were already discovered and exported by the sampled countries before 1995, indicating that MNCs operating in Malaysia have been progressively upgrading their product mix as manufacturing capability and local costs have advanced.

Comparing the sum of the productivity values for the "new" electronic and electrical products, Indonesia took the lead with the highest value of US$317,426. Indonesia had the largest number of electronics and electrical products categorized as new (21 compared with 11 for China, 2 for Thailand, and 1 each for Korea and Singapore). The sum of the productivity values of the new commodities for China was US$161,354. Indonesia's low level of technological progress before 1995 meant that it could discover and bring more electronics and electrical goods into the export market. Table 3.26 provides a detailed listing of these new electronics and electrical commodities at the six-digit level. Malaysia did not discover any new commodity during this time.

Finally, the sum of the productivity values for the cyclical group was highest for Indonesia again, at US$831,985. Malaysia was second at US$323,609. Thailand,

Table 3.26 New Electronic and Electricals That Appeared from 1995 to 2006 and Their 2000–04 PRODY

Country	Product	Product name	PRODY
China	902211	Medical x-ray apparatus	20109.1
China	854041	Magnetron tubes	17667.2
China	854049	Microwave tubes, except magnetron and klystron	17667.2
China	851782	Telegraphic apparatus, not elsewhere specified	17355.0
China	854219	Monolithic integrated circuits, except digital	16473.8
China	854220	Hybrid integrated circuits	16473.8
China	850611	Manganese dioxide primary cells or batteries, volume < 300 cc	12889.2
China	850612	Mercuric oxide primary cells or batteries, volume < 300 cc	12889.2
China	850613	Silver oxide primary cells or batteries, volume < 300 cc	12889.2
China	846921	Typewriters, electric, > 12 kg, nonautomatic	8470.0
China	846931	Typewriters, nonelectric, > 12 kg	8470.0
Indonesia	902221	Medical apparatus using alpha, beta, or gamma radiation	20109.1
Indonesia	902229	Nonmedical apparatus using alpha, beta, or gamma radiation	20109.1
Indonesia	902290	Parts and accessories for radiation apparatus	20109.1
Indonesia	854020	Television camera tubes and other photocathode tubes	17667.2
Indonesia	854042	Klystron tubes	17667.2
Indonesia	854081	Receiver or amplifier valves and tubes	17667.2
Indonesia	900911	Electrostatic photocopiers, direct process	17269.1
Indonesia	850133	DC motors, DC generators, output 75–375 kW	17025.1
Indonesia	850520	Electromagnetic couplings, clutches, and brakes	14631.8
Indonesia	853090	Electric signal, safety, and traffic controller parts	14631.8
Indonesia	854310	Particle accelerators	14631.8
Indonesia	851120	Ignition magnetos, magneto-generators, and flywheels	14572.4
Indonesia	841012	Hydraulic turbines, waterwheels, power 1,000–10,000 kW	13834.9
Indonesia	841231	Pneumatic power engines or motors, linear acting	13834.9
Indonesia	841280	Engines and motors, not elsewhere specified	13834.9
Indonesia	850612	Mercuric oxide primary cell, battery, volume < 300 cc	12889.2
Indonesia	850740	Nickel-iron electric accumulators	12889.2
Indonesia	850212	Generating sets, diesel, output 75–375 kVA	11781.9
Indonesia	850220	Generating sets, with spark ignition engines	11781.9
Indonesia	853940	Ultraviolet or infrared lamps, arc lamps	9511.6
Korea, Rep. of	841012	Hydraulic turbines, waterwheels, power 1,000–10,000 kW	13834.9
Philippines	847210	Office duplicating machines	17269.1

(*continued*)

Table 3.26 (*continued*)

Country	Product	Product name	PRODY
Philippines	853080	Electric signal, safety, and traffic controls, not elsewhere specified	14631.8
Philippines	853210	Fixed power capacitors (50/60 hertz circuits)	14631.8
Philippines	854520	Carbon and graphite brushes	14631.8
Philippines	854590	Battery carbons and carbon electrical items, not elsewhere specified	14631.8
Philippines	851290	Parts of cycle and vehicle light, signal, and similar equipment	14572.4
Philippines	850152	Air-conditioning motors, multiphase, output 0.75–75 kW	11781.9
Philippines	853921	Filament lamps, tungsten halogen	9511.6
Singapore	841012	Hydraulic turbines, waterwheels, power 1,000–10,000 kW	13834.9
Thailand	854020	Television camera tubes and other photocathode tubes	17667.2
Thailand	852210	Pickup cartridges	15707.7

Source: UN Comtrade.

India, and Korea followed at US$302,020, US$279,325, and US$278,081, respectively. Taiwan, China; China; Singapore; and Japan were at the lower end of this distribution with values of US$222,909, US$142,398, US$109,115, and US$64,329, respectively. Again, the number of cyclical electronics and electricals was highest in Indonesia at 54 compared with 20 in Malaysia and Thailand; 19 in Korea; 18 in India; and 13 in Taiwan, China. Hence, using the productivity values established in earlier studies, Malaysia's technological achievements can be ranked relatively high, although most products are manufactured by MNCs, and sophisticated components are mostly imported.

Export growth can also correlate with trends in the diversification of a country's exports toward faster-growing markets that exploit the potential of the extensive margin (Brenton and Newfarmer 2007). Compared with 1995, Malaysia exported more to China in 2007, which is a positive development (see table 3.27). However, the top three destinations—the United States, Singapore, and Japan—remained unchanged, although in 2007 they absorbed 39 percent of exports as against nearly 54 percent in 1995. Intraindustry and intraregional trade has been on the rise in East Asia for the past decade, and quite possibly this trend might continue. However, as noted previously, greater geographic concentration of manufacturing encouraged by increasing transport costs and the MNCs' preference for fewer suppliers might slow or reverse the increase, unless Malaysian firms diversify and upgrade their product offerings, thereby inducing MNCs to source more of their inputs from Malaysia.

The Evolution of Revealed Comparative Advantage

To help pull things together with a broad indication of industrial strengths based on exports, the study computed the revealed comparative advantage (RCA) using

Table 3.27 Destination of Malaysia's Exports, 1995, 2000, and 2007

	Share of exports received (%)									
Year	United States	Singapore	Japan	Hong Kong, China	United Kingdom	Thailand	Germany	Taiwan, China	Korea, Rep. of	China
1995	20.7	20.3	12.7	5.4	4.1	3.9	3.2	3.1	2.8	2.7
2000	20.5	18.4	13.1	4.5	3.1	3.6	2.5	3.8	3.3	3.1
2007	15.6	14.6	9.1	4.6	1.6	5.0	2.5	2.7	3.8	8.8

Source: UN Comtrade.
Note: Order of economies reflects share of exports received in 1995.

Malaysia's export baskets for 1995, 2000, and 2007.[20] In 1995, more than half of the top 10 exports with the highest ratings were resource-based products, and none were in the high-tech category (see table 3.28). The remaining three were medium-tech products, including audio equipment and turntables. The two leading items were both wood products; the third was palm kernel oil.

Seven years later, wood products had retained the two top spots, followed by thorium ores and palm oil. In all, there were five resource-based products and three high-tech items led by cinematographic cameras (see table 3.29). In 2007,

---

[20] The RCA is often used to measure the export competitiveness of a commodity (or an industry) of a country. It is a ratio of two shares: (a) the share of a commodity's export in the overall exports of a country and (b) the share of the same commodity in global exports:

$$RCA = \frac{Export_{ij} / \sum_i Export_{ij}}{\sum_j Export_{ij} / \sum_i \sum_j Export_{ij}},$$

where i denotes the commodity and j denotes countries (over the set of commodities, $i = 1 \ldots I$, and over the set of countries, $j = 1 \ldots J$). An RCA greater than 1 means that the country has a revealed comparative advantage in that commodity, assuming that the numerator and the denominator are increasing. If that is not the case, greater care is needed in interpreting the results. Rearranging the equation, one can obtain

$$RCA = \frac{Export_{ij} / \sum_j Export_{ij}}{\sum_i Export_{ij} / \sum_i \sum_j Export_{ij}}.$$

The numerator is now country j's market share of commodity i in the world export market, and the denominator is country j's share of exports in overall world exports. Thus, even if country j is losing market share, if overall exports from country j relative to world exports are shrinking faster, the RCA will be greater than 1 (Lall, Weiss, and Zhang 2006).

Table 3.28 Top 10 Commodities with Highest Revealed Comparative Advantage in Malaysia, 1995

Product	Product name	RCA	PRODY	Technology classification
440332	Logs, white lauan, meranti, seraya, yellow meranti, alan	57.03	2,287	RB1
440331	Logs, meranti (light or dark red), bakau	56.07	2,287	RB1
151329	Palm kernel and babassu oil, fractions, simply refined	52.65	4,661	RB1
851931	Turntables with automatic record-changing mechanism	50.91	15,997	MT3
440333	Logs, keruing, ramin, kapur, teak, jongkong, merbau, and so forth	50.27	2,287	RB1
151190	Palm oil or fractions, simply refined	48.68	4,635	RB1
440721	Lumber, meranti red, meranti bakau, white lauan, and so forth	45.86	3,667	RB1
851921	Record players without built-in loudspeaker, not elsewhere specified	44.03	15,997	MT3
261220	Thorium ores and concentrates	40.78	13,865	RB2
851939	Turntables, without record changers	38.43	15,997	MT3

Source: UN Comtrade.
Note: MT3 = engineering products; RB1 = agro-based products; RB2 = other resource-based products. Technology classification is based on Lall (2000).

Table 3.29 Top 10 Commodities with Highest Revealed Comparative Advantage in Malaysia, 2000

Product	Product name	RCA	PRODY	Technology classification
440331	Logs, meranti (light or dark red), bakau	57.36	2,287	RB1
440721	Lumber, meranti red, meranti bakau, white lauan, and so forth	44.31	3,667	RB1
261220	Thorium ores and concentrates	40.85	13,865	RB2
151190	Palm oil or fractions, simply refined	40.53	4,635	RB1
851931	Turntables with automatic record-changing mechanism	39.74	15,997	MT3
151329	Palm kernel and babassu oil, fractions, simply refined	39.04	4,661	RB1
900711	Cinematographic cameras for film <16 mm	38.55	12,554	HT2
900719	Cinematographic cameras for film >16 mm	34.71	12,554	HT2
903130	Profile projectors, not elsewhere specified	32.89	21,451	HT2
230660	Palm nut or kernel oil cake and other solid residues	32.83	5,718	PP

Source: UN Comtrade.
Note: HT2 = other high-technology products; MT3 = engineering products; PP = primary products; RB1 = agro-based products; RB2 = other resource-based products. Technology classification is based on Lall (2000).

Table 3.30 Top 10 Commodities with Highest Revealed Comparative Advantage in Malaysia, 2007

Product	Product name	RCA	PRODY	Technology classification
900921	Other photocopying apparatus, incorporating an optical system	71.32	17,269	HT1
551432	Yarns of different colors, 3-thread or 4-thread twill, including cross-twill, of polyester staple fibers	71.19	12,873	MT2
851929	Other record players	70.27	15,330	MT3
851991	Cassette players, nonrecording	69.27	15,330	MT3
851910	Coin- or disk-operated record players	68.97	15,330	MT3
440331	Logs, meranti (light or dark red), bakau	68.80	5,595	RB1
330126	Essential oils other than those of citrus fruit, of vetiver	68.63	13,858	RB2
261220	Thorium ores and concentrates	68.49	13,865	RB2
851921	Other record players, without loudspeaker	68.19	15,997	MT3
441121	Fiberboard, a density exceeding 0.5 g/cm³ but not exceeding 0.8 g/cm³, not mechanically worked or surface covered	63.30	11,754	RB1

Source: UN Comtrade.
Note: HT1 = electronic and electrical products; MT2 = process industry; MT3 = engineering products; RB1 = agro-based products; RB2 = other resource-based products. Technology classification is based on Lall (2000).

photocopying apparatus was at the top, followed by medium-tech products, displacing resource-based items, although these products still remain in the top 10 (see table 3.30). Overall, the impression conveyed by the diverse mix of products is that Malaysia's apparent comparative advantage continues to lie in resource-based products, mainly in wood and palm oil. By comparison, the RCA for Singapore is more solidly in high-tech electronic and engineering products and in chemical compounds (see appendix table A.6). The RCA for Korea in 2007 mainly points to medium-tech products, including ships, chemicals, electronic products, and yarn (see appendix table A.4). Thailand's RCA for 2007 is heavily slanted toward resource-based and agricultural products (see appendix table A.8).

Table 3.31 lists the changes in RCA from 1995 to 2007 for Malaysia's 10 largest export commodities. Six commodities are classified as high-tech. RCAs for these commodities are mostly increasing, although slowly. Clearly, Malaysia enjoys a comparative advantage in palm oil, and the export value is significant. For now, Malaysia is also maintaining its comparative advantages in the other commodities.

In sum, the RCA measure helps make future choices somewhat clearer, showing in line with the other indicators that the manufacturing capabilities of Malaysian firms are strongest in medium-tech resource-based products. The country is a competitive producer of some electronic products, but they are mainly items

Table 3.31 Top 10 Malaysian Exports in 2007 and Their Revealed Comparative Advantages in 1995–2007

Product	Product name	RCA, 1995	RCA, 2000	RCA, 2007	Trade value, 2007 (US$ million)	Technology classification
854211	Monolithic integrated circuits, digital	0.24	2.64	4.13	11,000	HT1
847330	Parts and accessories for machines of heading 84.71	3.62	5.38	5.36	10,400	HT1
270900	Petroleum oils and oils obtained from bituminous minerals, crude	1.47	0.79	1.11	9,760	PP
271111	Liquefied natural gas	10.53	10.11	10.87	7,620	PP
847120	Digital computers with central processing and input-output units	0.25	1.65	4.93	7,170	HT1
151190	Palm oil or fractions, simply refined	50.28	41.43	38.12	6,980	RB1
854220	Hybrid integrated circuits	12.92	4.14	21.09	6,210	HT1
271000	Petroleum oils and oils obtained from bituminous minerals, other than crude; preparations not elsewhere specified or included, containing by weight 70% or more of petroleum oils or of oils obtained from bituminous minerals			0.85	5,770	RB2
854219	Monolithic digital integrated circuits, other, including circuits obtained by a combination of bipolar and metal oxide semiconductor technologies	5.36	4.94	2.44	4,420	HT1
847199	Automatic data processing machines and units, not specified elsewhere	1.08	3.42	7.64	4,140	HT1

Source: UN Comtrade.
Note: HT1 = electronic and electrical products; PP = primary products; RB1 = agro-based products; RB2 = other resource-based products. Technology classification is based on Lall (2000).

assembled by MNCs and do not reflect the technological expertise of Malaysian firms.[21]

Opportunities Defined by Mapping Product Spaces

The burning question for Malaysia and other Southeast Asian countries is all about the next generation of products. Given current comparative advantage, how can one identify the options for diversifying the industrial mix of tradable products? This question takes us into the realm of speculation, guided by a useful technique for mapping "product spaces," pioneered by Hausmann and Klinger (2006). Their

---

[21] An index of the sophistication of exports by country constructed by Lall, Weiss, and Zhang (2006) on the basis of the classification derived by Lall (2000) shows that in 1990 Malaysia ranked 14th with a score of 68 (the United States ranked first with a score of 84). In 2000, Malaysia's score was 63, behind the Philippines (64) but ahead of Thailand (62).

approach assumes that each commodity produced gives rise to different opportunities for future diversification. That is, some products offer easier and multiple diversification paths to other related products, whereas others do not. In general, primary and resource-based products do not lead to many opportunities for diversification. By contrast, manufacturing goods, such as electronics, generate skills and assets that are similar to those required for the production of other manufacturing commodities and hence are classified as high-value products.

The product space mapping technique helps a country identify the potential diversification opportunities arising from each of its exports. The density of each commodity gives the probability that a country will export a pair of goods conditional on its already exporting at least one of the goods. The more a country specializes in high-value goods (with regard to highest densities), the greater is its potential for diversification into other high-value products.

Graphical representations of the product spaces for Indonesia, Malaysia, the Philippines, and Thailand are presented in figure 3.20 and for China in figure 3.21. They show clear differences in the pattern of exports and the potential for diversification in each of these countries relative to Malaysia. The period for this analysis is 2000 to 2004. The x axis is the inverse of the density (that is, closer to the origin indicates higher density); the y axis measures the difference between PRODY and EXPY[22] (that is, a positive number means "upgrading" in a sense of exporting more sophisticated commodities relative to the overall export basket).

In the case of China (see figure 3.21), the commodities that are in the area of high density are mostly higher-valued commodities, such as engineering and high-technology goods.[23] Appendix table B.1 shows that the "upscale" commodities with the highest densities are mostly in the engineering, electronic and electrical, light manufacture, and processed products. In contrast, for Malaysia, fewer commodities are in the denser part of the product space, mainly those in electronics and electricals, most of which are assembled by MNCs with some inputs from local firms (see figure 3.20). The highest densities for upscale products in Malaysia ranged from 0.27 to 0.21 (see appendix table B.6), whereas for China they ranged from 0.54 to 0.46. Clearly, firms in Malaysia have the opportunity to diversify into new products, especially in electronics (and some automotive, processed, and engineering products), but diversification would entail investment in product design and development and in research and development so as to strengthen absorptive capacity. In view of the large role of MNCs in Malaysia and the absence of significant local original design manufacturers and contract manufacturers, the likelihood that firms in Malaysia can readily diversify into new and more sophisticated products is smaller relative to China.

---

[22] EXPY is calculated as a weighted sum of PRODY, signifying the sophistication of the export basket of a country.

[23] The discussion in this section focuses on the "upscale" goods—that is, PRODY and EXPY are both positive.

Figure 3.20 Product Space of Selected Southeast Asian Countries, 2000–04

a. Indonesia

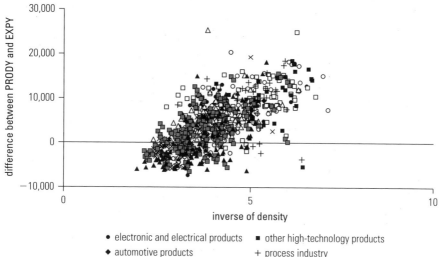

- electronic and electrical products
- automotive products
- engineering products
- other low-technology products
- other resource-based products
- other high-technology products
- process industry
- textiles, garments, and footwear
- agro-based products
- primary products

b. Malaysia

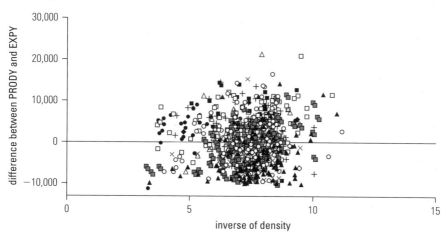

- electronic and electrical products
- automotive products
- engineering products
- other low-technology products
- other resource-based products
- other high-technology products
- process industry
- textiles, garments, and footwear
- agro-based products
- primary products

Figure 3.20 (*continued*)

c. Philippines

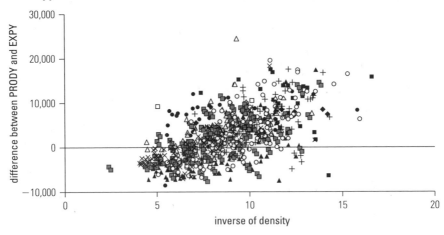

- electronic and electrical products
- automotive products
- engineering products
- other low-technology products
- other resource-based products
- other high-technology products
- process industry
- textiles, garments, and footwear
- agro-based products
- primary products

d. Thailand

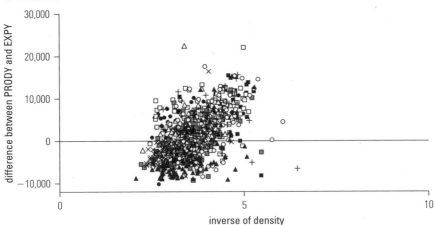

- electronic and electrical products
- automotive products
- engineering products
- other low-technology products
- other resource-based products
- other high-technology products
- process industry
- textiles, garments, and footwear
- agro-based products
- primary products

Source: Authors' calculations.

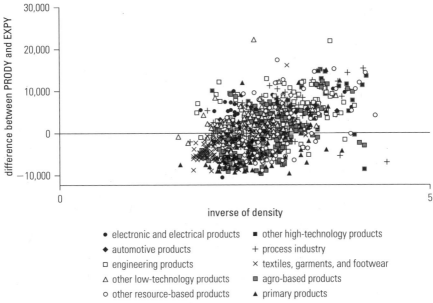

Figure 3.21 Product Space of China, 2000–04

- electronic and electrical products
- automotive products
- engineering products
- other low-technology products
- other resource-based products
- other high-technology products
- process industry
- textiles, garments, and footwear
- agro-based products
- primary products

Source: Authors' calculations.

Appendix table B.6 also identifies the SITC2 four-digit-level products into which Malaysia can most easily diversify. In particular, consumer electronics consisting of products such as televisions, dictating machines, and other sound recorders and reproducers comprise the low-hanging fruit for Malaysian manufacturers; however, given rising labor costs and saturated world markets, these are less attractive. Three products from the electronics and electrical industry follow next: (a) electronic microcircuits; (b) peripheral units, including control and adapting units; and (c) diodes, transistors, and similar semiconductor devices. The other promising commodities for the engineering industry are radios, printed circuits and parts, and radios for motor vehicles—product groups also subject to strong competitive pressures. In the electronics and electrical industry, the most promising product categories are complete digital data processing machines; off-line data processing equipment; parts of and accessories suitable for electronic calculating and accounting and other automatic data processes; other electrical valves and tubes; parts of electrical apparatus for line telephones, microphones, loudspeakers, amplifiers, and aerial reflectors of all kinds; television picture tubes, cathode ray tubes; calculating machines and other accounting machines; and other electrical machinery and equipment. Two commodities at the SITC2 four-digit level in medium-technology processes with high densities are (a) polycarboxylic acids and their anhydrides and (b) polystyrene and its copolymers. The only subsectors in the automotive industry that have relatively high densities (0.21) are (a) wheelchairs, both motorized and nonmotorized, and (b) parts and accessories for motorcycles, autocycles, and bicycles.

Thailand concentrates on higher-value products relative to Malaysia. The commodities with the highest densities range from 0.40 to 0.35 (see appendix table B.9), well above the range computed for Malaysia (see figure 3.20), with an emphasis on electronics and electricals, engineering, processed products, and some low-tech manufactures.

Indonesia lags Malaysia in technological capabilities. Figure 3.20 and appendix table B.4 show that Indonesia's highest-density upscale products were primarily in the resource-based and primary products sectors, with some in light manufacturing (and a couple in engineering).[24]

Figure 3.20 shows that the Philippines also has a lower "open forest" (that is, upscale product space) than Malaysia. Its highest-density products are in the 0.23 to 0.16 range (see appendix table B.7). The prominence of resource-based industries (with some specialization in engineering and electronics and electrical goods) is also notable.

Similar estimates for Hong Kong, China; Korea; and Singapore (not shown here) reveal that the open forest of Hong Kong, China, is denser than that of Malaysia. Its highest-density products ranged from 0.40 to 0.31 (appendix table B.2), much higher than Malaysia's. The focus of Hong Kong, China, was on high-value commodities—namely, engineering, electronics and electricals, other high-technology goods, processed and automotive products, and light manufactures. Korea's densest forest contains mainly high-value products. However, its top 20 densest commodities fall in the 0.41 to 0.32 range, which is much higher than the score for Malaysia (see appendix table B.5). Singapore specializes in commodities such as electronics and electricals, engineering, and processed materials (and very few resource-based goods; see appendix table B.8), similar in many respects to Malaysia. Surprisingly, its open forest is also less dense, ranging from 0.30 to 0.23.

An Overview of Export Capabilities

What we learn about Malaysia's capacity, potential, and future options from an analysis of export trends, pattern, and composition; from revealed comparative advantage; and from the mapping of product space is all fairly consistent.

---

[24]Coxhead and Li (2008) attribute this finding to the lower level of human capital investment in Indonesia relative to Malaysia. If Indonesia had a level of foreign direct investment and human capital similar to Malaysia's during the period from 1995 to 2005, the share of skill-intensive exports would account for 22 percent of the total (the actual was 13 percent during this period).

Malaysia's current strength is in resource-based manufacturing, processing, and assembly activities in line with its resource endowment, the past industrial allocation of foreign direct investment by MNCs, and weak domestic technological capability. Technological catch-up is proceeding slowly. The significance of the electronics sector, although well grounded in production and exports, does not reflect the true technological capacity of local firms. Malaysia's domestic inputs to the electronics sector are still relatively low tech. Its two great advantages lie in the depth of manufacturing expertise and the relatively low productivity-adjusted cost of labor. Improving the technical and communication skills of workers and strengthening the innovation capabilities of firms are needed to speed up the technological catch-up process. Consolidating comparative advantage in those activities that are within reach would seem to be the wisest strategy. This effort could be combined with steady investment in selected high-tech areas commensurate with the country's capacity to effectively absorb and evolve the relevant technologies.

The Scope for Expanding Tradable Services

Much has been made of the growth and export potential of tradable services. India's success in capturing the markets for outsourced information and communication technology, design, engineering, and some medical services has fueled the ambitions of Southeast Asian countries, several of which are now looking beyond tourism and finance to a wider range of possibilities, including medical services to travelers. In Malaysia, as in Thailand, travel-related services are still growing faster than others (see figure 3.22). Transport services increased much more modestly. The available data for "other services" are from 1999 to 2007. Construction and computer and information exports are on the rising trend. The export performance of other services has been relatively weak, most conspicuously financial services, which, while fluctuating from year to year, remained virtually flat between 1999 and 2007 (see figure 3.23). Relative to other East Asian comparators, Malaysia's service exports are performing comparably to those of Thailand and ahead of those of Indonesia and the Philippines (see figure 3.24).

Nevertheless, policy makers and suppliers of services are hopeful that Malaysia will be able to expand the exports of five kinds of services. First is tourism, by attracting more tourists from China, Europe, and the Middle East. Attempts to achieve this goal are ongoing and include international marketing efforts; investment to restore cultural heritage sites (in Penang, for example); and improvement of the tourist infrastructure in coastal centers (such as Langkawi). However, to raise the daily expenditure by visitors and induce them to extend their stay will require the development of ecotourism, other recreational activities, more sites for tourists to visit, and a periodic refreshing of the recreational options. The Malaysia My Second Home program is also being actively promoted to attract long-staying visitors.

Figure 3.22 Composition of Service Exports from Malaysia, 1995–2007

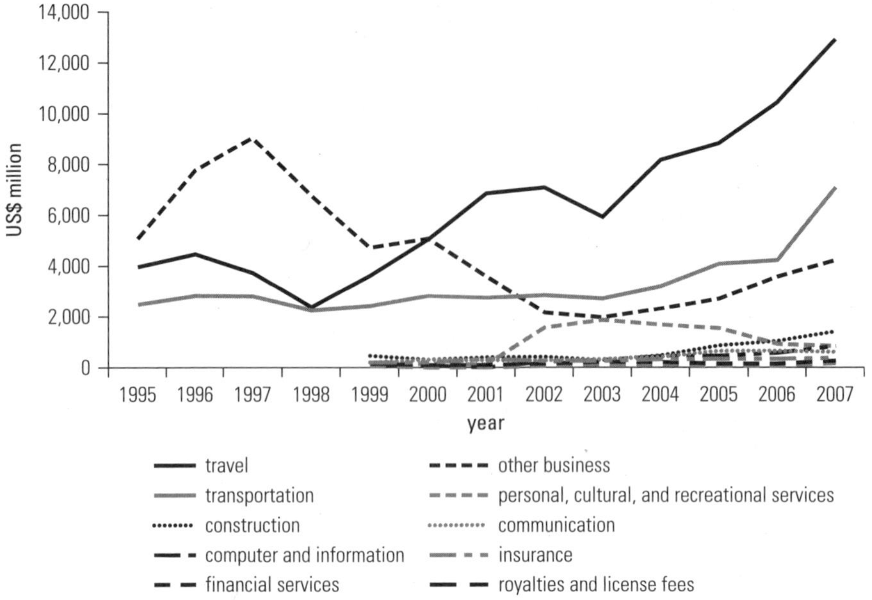

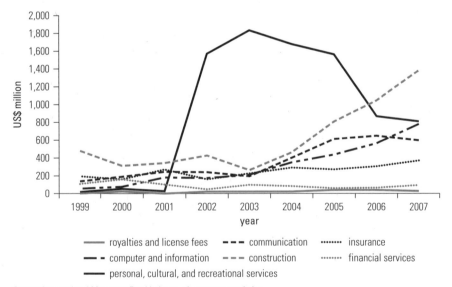

Source: International Monetary Fund balance of payments statistics.

Figure 3.23 Composition of Selected Service Exports from Malaysia, 1999–2007

Source: International Monetary Fund balance of payments statistics.

Figure 3.24 Service Exports from Southeast Asian Countries, 1995–2007

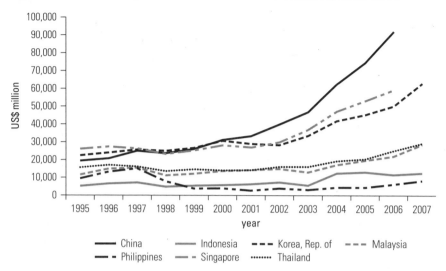

Source: International Monetary Fund balance of payments statistics.

Second, provision of medical services is an expanding industry in Penang, catering mainly to visitors from Sumatra, Indonesia, the vast majority from the Chinese community in Medan and its environs, because that area has strong ethnic and linguistic ties with the ethnic Chinese community in Penang (Hokkien). Almost 200,000 visitors used Penang's seven international hospitals in 2006 for everything from routine physical exams to heart-related procedures and hip and knee replacements.[25] Each visitor spent approximately two and one-half days in a hospital (not including any additional time spent in Penang or other resorts to convalesce) and was accompanied by at least one additional person who generated revenues for local merchants, hotels, and owners of long-term rental accommodations. Should this industry continue to expand and generate spillovers for local providers of inputs, it could become an important source of growth. But projecting future flows of medical tourists and localized linkage effects is difficult because Malaysia is facing strong competition from India, Singapore, and Thailand (with countries such as Mauritius also attempting to obtain some of the business) for medical tourists sourced from the region and

---

[25]Two of these hospitals now have Joint Commission International (JCI) certification. There are three JCI-accredited hospitals in Thailand. Those in Thailand account for more than half of the foreign patients, estimated to be more than 1 million a year. By 2015, it is projected that more than 7 million patients will visit Thailand for various medical treatments, assuming that the country remains on a stable development path (Smith, Chanda, and Tangcharoensathien 2009).

Table 3.32 Costs of Selected Procedures in Selected Countries

Procedure	Cost (US$)				
	India	Singapore	Thailand	United States	United Kingdom
Heart bypass graft surgery	6,000	10,417	7,894	23,938	19,700
Heart-valve replacement	8,000	12,500	10,000	200,000	90,000
Angioplasty	11,000	13,000	13,000	31,000–70,000	—
Hip replacement	9,000	12,000	12,000	22,000–53,000	—
Hysterectomy	—	13,000	10,000	—	—
Bone marrow transplant	300,000	—	—	250,000–400,000	150,000
Liver transplant	40,000–69,000	—	—	300,000–500,000	200,000
Neurosurgery	800	—	—	29,000	—
Knee surgery	2,000–4,500	—	8,000	16,000–20,000	12,000
Cosmetic surgery	2,000	—	3,500	20,000	10,000

Source: Smith, Chanda, and Tangcharoensathien 2009.
Note: — = not available.

from the Middle East.[26] These countries offer some of the procedures at much lower costs than the United Kingdom and the United States (see table 3.32). Improving medical services in Indonesia could also dampen tourist arrivals in Malaysia, as will slower growth in the region. Currently, the hospital industry in Penang has few links with suppliers of medical equipment and suppliers in the immediate vicinity.[27] Cultivating such links will call for proactive efforts by hospitals and public agencies, as well as entrepreneurial initiative on the part of manufacturers, suppliers of information technology services, and others to maximize the domestic value added. The supply of nurses and specialists will remain a constraint because of continuing emigration, particularly of nurses whose professional and English-language skills are much in demand in the Middle East. Emigration also would be moderated by weaker external demand for emigrants during 2009 to 2011.

A third tradable service, Islamic finance, has the potential to generate substantial turnover, and it has weathered the 2008 financial crisis better than other

---

[26] Relative to other service sectors, fewer World Trade Organization member countries are committed to opening health and education-related sectors compared with finance and telecommunications. The Indian medical transnational firm Wockhardt is a major competitor, as are Apollo Hospitals and Fortis Healthcare. Singaporean firms, such as Raffles Hospital and Parkway Group Healthcare, have established marketing offices in China, the Russian Federation, the Middle East, Southeast Asia, and South Asia to attract customers in these areas (Smith, Chanda, and Tangcharoensathien 2009).

[27] The hospital industry in Thailand also has forged few links with local suppliers of inputs.

segments of the financial sector. Malaysia's financial sector is reasonably well developed relative to that of Indonesia, for example, but there is mounting competition from Singapore, the Middle East, and Western providers, all of which are tailoring products to attract investors by offering special preferences. A plan to create an offshore banking enclave in Penang is in the air but is far from fruition. The slow growth of financial services since 2000 and the uncertainty generated by the financial crisis of 2008 and 2009 and what the crisis has revealed about the financial sector's contribution to growth argue against an overly optimistic assessment of the prospects.

Halal certification, which indicates that products are permissible under Islamic law, is the fourth expanding industry, with Malaysia vying to become a center for certifying foodstuffs, cosmetics, and medical products, among others. Again, the net earnings from this activity are uncertain, as is the scope for expanding the scale of certification activities internationally and generating substantial revenues. Malaysian entities are aggressively attempting to make the country a hub of the Halal certification business by hosting exhibitions and sensitizing Muslim consumers to the desirability of obtaining such an assurance.

Logistics services are a fifth source of export revenue. Malaysia is investing heavily in both surface and air transport-related infrastructure, thus far with limited results.

Malaysia has three main multilayered logistics hubs: (a) the Penang Logistics Platform for the Northern Corridor Economic Region, (b) the Greater Klang Valley Logistics Platform for the Central Economic Corridor, and (c) the Iskandar Logistics Platform for the Southern Development Region. Port Klang, with its free trade zone, is the focus of the government's efforts to develop a regional distribution center. The Iskandar Logistics Platform, encompassing three ports (including Tanjung Pelepas), and Senai International Airport also have considerable potential. Penang Port competes with ports in Singapore, Tanjung Pelepas, and the Klang Valley. Penang Port is also handicapped by its inability to handle large vessels, the low frequency of shipping services to the EU and United States, its reliance on coastal shipping, and its smaller industrial hinterland. In contrast, Penang's airport is a busy regional center serving the local electronics and semiconductor firms.

Frost & Sullivan's (2005) study of hub potential, which ranked countries by their attractiveness for logistics companies with reference to wage costs, quality of physical infrastructure, political stability, and business environment, put Malaysia in fourth place after China and Hong Kong, China; the Philippines; and Singapore. However, Kuala Lumpur's Port Klang, Malaysia's principal logistics center, still trails others in the region in the transport and communications rankings. In the Asia-Pacific region, only two full-fledged logistics platforms are included in the world's top 25 ports: Tokyo-Yokohama and Hong Kong, China. Other aspirants are Taipei-Keelung, Seoul-Incheon, Beijing-Tianjin, Shanghai-Ningbo, and Singapore. Kuala Lumpur still has much ground to cover before it can equal Singapore in terms of port and airport capabilities and international connectivity.[28] MNCs looking for a centralized distribution center outside Northeast Asia prefer Singapore to other locations.

Malaysia's logistics capacity is improving, and its global network connectivity position has risen from 28th in 2000 to 18th in 2004. However, its three principal logistics platforms remain suboptimal with regard to the integration of physical and information technology infrastructure, quality software services, and the presence of numerous world-class logistics suppliers. Furthermore, none of the Malaysian hubs is on a path to becoming a supply-chain nerve center that is a source of innovative sourcing and procurement solutions, data hosting, and reverse logistics. Malaysian hubs have access to land, which Singapore does not. Hence, the potential for growth is greater, but it will call for a carefully articulated strategy. Development of the Kuala Lumpur–Port Klang hub in the near term might have the highest payoff. There would also seem to be advantages to developing the Iskandar Platform in close coordination with Singapore. An approach that spreads resources widely and thinly may be less effective, especially in the face of strong competitive pressure from other logistics platforms in southern China, the Philippines, and Thailand.

---

[28]Where Port Klang outscores other ports in the region is in its low door-to-door total logistics costs to Europe, the Middle East, North America, and South Asia.

4

Imports and Foreign Direct Investment
Competition and Technology Transfer

Imports can contribute to growth through at least two channels.[1] They can force domestic producers either (a) to improve their production efficiency and the quality of their products and be innovative or (b) to exit from the business, a process that can reallocate resources to more productive uses. Imports also are helpful in identifying domestic opportunities for local firms willing to compete (Lawrence and Weinstein 1999). Technological advances embodied in imports of machinery and equipment provide a second channel.

The Malaysian economy is relatively open to imports if measured by tariff levels. Except for a few product categories, Malaysia's tariffs are low and comparable to those of its neighbors.[2] Figure 4.1 shows that from 1995 to 2007, nominal tariffs on total imports of Malaysia were below those of other comparators except Japan and Singapore. On imports of machinery and equipment, Malaysia's tariffs were in the 3 percent range and only those of Japan and Singapore were appreciably lower (see figure 4.2).

Although nominal tariffs may no longer be an issue, nontariff barriers and other restrictions have reduced competition in the domestic market. This situation is apparent from the low productivity, the lack of innovativeness, and the small size of most firms in the auto parts subsector, for example. These firms have been able to survive because of a variety of controls and contracts from public

---

[1] Mahadevan (2007a), using Granger causality tests, shows that imports lead gross domestic product growth in Malaysia and notes that imports are a source of embodied technical change. Marwah and Tavakoli (2004) find that net foreign direct investment inflow and imports contributed 0.53 percentage point to growth.

[2] This is not to say that trading partners do not have any complaints about the tariff structure of Malaysia. The most frequently voiced concern is the number of unbound tariff items (Ramasamy and Yeung 2007).

Figure 4.1 Average Tariffs on Total Imports, 1995–2007

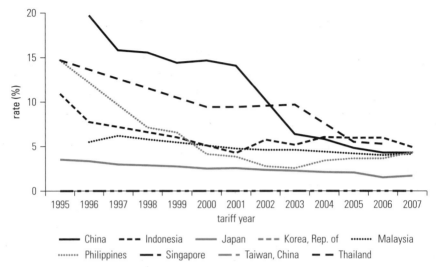

Source: United Nations Conference on Trade and Development Trade Analysis and Information System (UNCTAD-TRAINS).

Figure 4.2 Average Tariffs on Machinery and Equipment Imports, 1995–2007

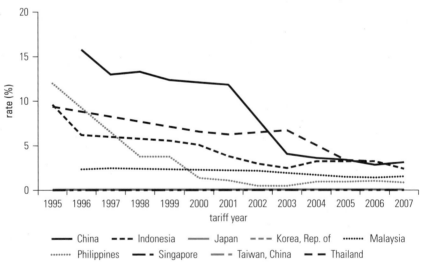

Source: UNCTAD-TRAINS.

entities, whereas under competitive conditions, mergers and exits would have resulted in a leaner and more robust industry.[3] One factor to be noted here is the absence of a competition law in Malaysia (Lee 2007). In addition, strategic industrial policies to assist local suppliers are also negating some of the apparent market openness suggested by the tariff regime.

Imports and Technology Transfer

A more tangible and immediate source of gains in productivity is the imports of machinery and equipment embodying new technologies that permit Malaysian producers to raise process efficiency and quality (see footnote 3 of chapter 2). None of this is news. Malaysian producers are generally experienced and knowledgeable about technologies available from overseas, helped by the presence of multinational corporations (MNCs), which are a valuable conduit for information. What is not happening in Malaysia is product and process development, which would give local producers an edge over the competition. Although the returns from investment in research and development (R&D) are frequently emphasized, a country can arguably, in the early or mid catch-up phase, derive more benefit from R&D conducted in advanced countries. By making greater efforts to assimilate the findings from international R&D, a country can complement and magnify the gains from domestic R&D. It can also partially circumvent the large costs of development and testing. Acharya and Keller (2007: 3) find that the "contribution of international technology transfer often exceeds the effect of domestic R&D on productivity. On average, the combined impact of R&D investments in six countries close to the technology frontier, the U.S., Japan, Germany, France, the UK, and Canada, is about three times as large as that of domestic R&D. Moreover, the global patterns of technology transfer are highly asymmetric ... some countries benefit more from foreign technology than others."[4] They benefit more from R&D in some countries than in others because of differences in absorptive capacity. Technology transfer from some countries, such as the United Kingdom and the United States, is more through trade, whereas that from Germany and Japan is through nontrade channels. Furthermore, the importance of technology transfer has risen since the 1990s. Three indicators of embodied technology are (a) the volume of investment in

---

[3]Several studies conducted by the McKinsey Global Institute and summarized by Palmade (2005) indicate that in countries that follow sound macroeconomic policies, the growth of productivity is frequently being restrained by microeconomic distortions associated, for instance, with land use, labor laws, rent restrictions, price controls, measures limiting the exit of small firms, and other discriminatory practices.

[4]See Aghion and Howitt (2005), who observe that returns from R&D spending are greater the closer a country is to the technology frontier.

Figure 4.3 Composition of Machinery and Equipment Imports by Malaysia, 1995–2007

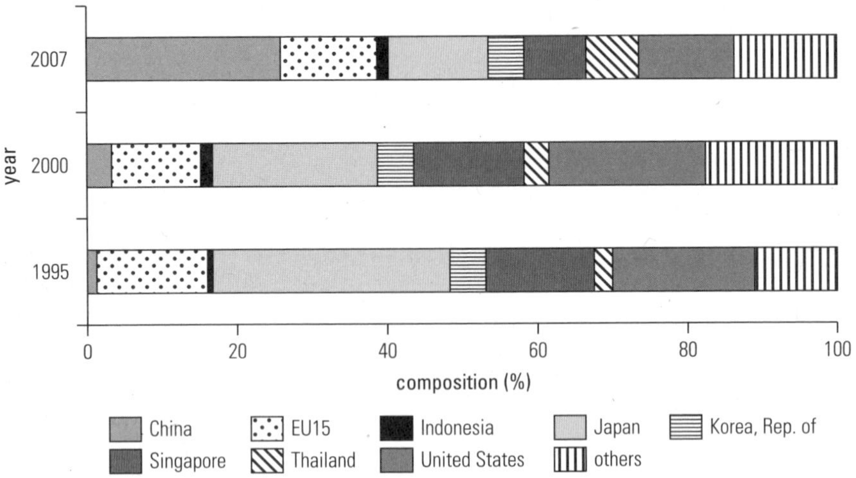

[Bar chart showing composition (%) for years 1995, 2000, and 2007, with categories: China, EU15, Indonesia, Japan, Korea, Rep. of, Singapore, Thailand, United States, others.]

Source: United Nations Commodity Trade Statistics Database (UN Comtrade).
Note: Machinery and equipment imports are from category 84 (excluding parts), and soldering and welding machines are from category 85. The EU15 are the 15 members of the European Union prior to its expansion on May 1, 2004: Austria, Belgium, Denmark, Finland, France, Germany, Greece, Ireland, Italy, Luxembourg, the Netherlands, Portugal, Spain, Sweden, and the United Kingdom.

fixed productive assets, (b) the import of capital equipment, and (c) the countries from which plant and equipment are purchased. Gross domestic investment in Malaysia dropped from 43.6 percent of gross domestic product (GDP) in 1995 to 20.7 percent of GDP in 2006;[5] however, imports of finished capital goods (not including transport equipment), which had reached US$6.4 billion in 2000, were valued at US$9.4 billion (in nominal terms) in 2006, a respectable increase. But the sources of imports of plant and equipment have changed, with less being imported from Japan and an increasing share coming from China and Thailand, whereas imports from the European Union, the Republic of Korea, Singapore, and the United States fluctuate (see figure 4.3). It is difficult to infer from the data on imports whether they are more or less technology intensive.[6] For instance, some of the imports might be from Japanese transplants in China, but to determine whether they are would require an item-by-item scrutiny of

---

[5] It diminished further in 2008–09 (Tan and Ahya 2009).
[6] Although the assumption is that capital equipment exports from advanced countries are of higher quality, some research has yielded findings to the contrary. A study has shown that developing countries tend to import lower-quality equipment than advanced countries (Navaretti and Soloaga 2001). For a review of capital good imports as a source of technology transfer, see Hoekman and Javorcik (2006) and Nabeshima (2004).

customs invoices for the equipment imported. The aggregate data available suggest that technology transfer embodied in capital imports has probably diminished. Very likely, producers in Malaysia are switching their orders to cheaper sources in China. Korean and Chinese equipment could embody less sophisticated technology, which affects the upgrading of production processes and product quality. This tendency argues for greater attention toward building domestic innovation capacity.

Malaysia may now be on the threshold of the stage where the entry, competitiveness, and growth of domestic firms hinge on the firms' own R&D efforts supplemented by more basic research in universities and specialized institutions. Technology transfer from imports of mature products and processes is likely to be smaller than from items that are at an early stage of their lifecycle. As noted earlier, Malaysian producers need to transition from commodities to product categories that are at an early stage of their lifecycle, when the room for innovation is greatest and there is scope for earning large rents. Adaptation and incremental improvement of imported technologies and the domestic production of import-substituting products, which add value locally, depend more on domestic R&D capital and on innovation capability. Foreign R&D is a partial substitute for domestic R&D. As a country moves closer to the technology frontier, the role of domestic R&D increases, as do the returns from research. How readily local suppliers are able to substitute for imports is related to their design and engineering skills, but that is not all. Domestic importers generally prefer local suppliers, and how actively such importers help potential local suppliers to find and acquire technologies that enable them to compete against foreign producers enhances domestic absorptivity, as noted by Blalock and Veloso (2007).[7] Their research on Indonesian firms also shows that the productivity effect of technology transfer is "greater in supply industries with high firm concentration. Firms in these concentrated industries are typically subject to less local competitive pressures and, therefore, exposure to international competition from imports will likely [lead to] greater marginal productivity gains" (Blalock and Veloso (2007: 1137–38).[8]

---

[7]Typically, the subsidiaries of MNCs have local vendor development programs to identify and train local suppliers for bulky or simple products that can be locally sourced. In some cases, such a program can evolve into a regional vendor development program that evaluates suppliers from surrounding countries. This scenario is one avenue of (vertical) technology transfer from MNCs. MNCs in Malaysia have engaged in such programs (Ariffin and Figueiredo 2004).

[8]Research on Mexico has yielded similar findings (Jordaan 2005). Technology transfer among narrowly defined industrial sectors (four-digit level) does not seem to be affected by distance. However, at the three-digit level, distance has a negative effect on technology transfer (Orlando 2004).

Figure 4.4 Net Inflows of FDI, 1995–2006

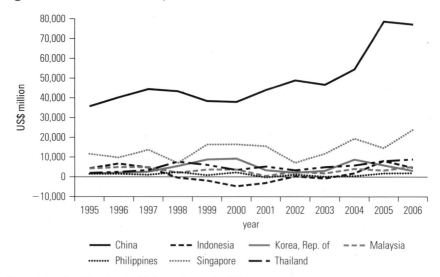

Source: International Monetary Fund balance of payments statistics.

Patterns of Foreign Direct Investment

Trends in exports and the pattern of export diversification suggest opportunities for future growth. However, these opportunities need to be validated by the distribution of investment across industries, which shows concretely how investors view the emerging prospects. Investment is arguably a more sensitive indicator of subsectoral opportunities as perceived by the foreign and local business communities, whose reading of the market possibilities marshals the highest level of expertise and draws on the largest fund of experience. Approved foreign direct investment (FDI) in Malaysia was US$5.51 billion in 2006. It rose to US$9.72 billion in 2007, a year when global FDI climbed to record levels, and diminished in 2008 and 2009, reflecting the massive downturn in international capital flows starting in the second half of 2008.[9] China receives the lion's share of FDI inflow to East Asia, followed by Singapore (see figure 4.4). Indonesia, the Philippines, and Thailand are in much the same situation as Malaysia, with FDI levels fluctuating modestly around a stable trend.

---

[9] Global FDI flow in 2008 declined by 15 percent from the record high of US$1.9 trillion in 2007. FDI flows are expected to decline in 2009 and 2010, and the recovery of FDI is not expected to start until 2011 (UNCTAD 2009).

Figure 4.5 Top 12 FDI Sectors in Malaysia, 1999–2003

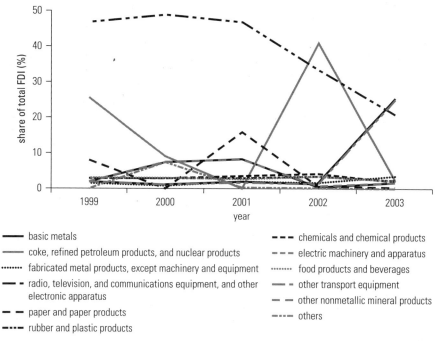

Source: Association of Southeast Asian Nations (ASEAN) FDI Database, 2004.

The data available on FDI by industrial subsectors are for projects approved between 1999 and 2003. A high percentage—up to 75 percent—of these projects were actually implemented, so the information is a reliable guide to market expectations of returns on investment over the medium term.[10] From figures 4.5 and 4.6, it is apparent that electronics products and telecommunications equipment have traditionally attracted the most FDI in terms of value,

---

[10]The rate of implementation has been especially high since 1998, when FDI in the manufacturing sector was liberalized to allow 100 percent foreign ownership without any export requirement. This development was welcome because foreign equity participation can improve the performance of firms (for the case of China, see Ran, Voon, and Li 2007). However, seven subsectors were exempted from this guideline (Tham 2004). Such restrictions still apply to many services sectors, which require either a Malaysian resident natural person or a joint venture with local firms (Ramasamy and Yeung 2007). Malaysia does not have any laws regarding FDI. All the regulations are implemented through the "Foreign Equity Guidelines," which permit the government to customize conditionality by individual project (Tham 2004). Korea also restricts FDI in services, except with some qualifications in financial services ("South Korea: FDI Key" 2004).

Figure 4.6 Top 12 FDI Sectors in Malaysia, 1999–2003

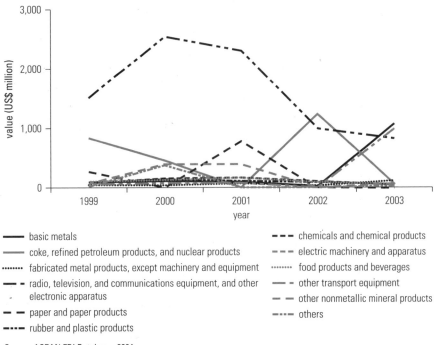

— basic metals
— coke, refined petroleum products, and nuclear products
········ fabricated metal products, except machinery and equipment
— · radio, television, and communications equipment, and other electronic apparatus
— — paper and paper products
—·—· rubber and plastic products
— — — chemicals and chemical products
— — — electric machinery and apparatus
········ food products and beverages
— · other transport equipment
— — other nonmetallic mineral products
—·—· others

Source: ASEAN FDI Database, 2004.

but also that total investment peaked in 2000 and has since been declining. Investment in other subsectors is mostly stable and fairly low. There are a couple of spikes, one being a surge of investment in the petroleum products industry in 2002, which then subsided. The second was a sudden increase in the investment in basic metals in 2003; this surge is also likely to have been temporary and needs to be verified from more recent data. Disaggregated data on the four major industrial regions in Malaysia largely reinforce these findings. Electronics and electrical products lead the field. The picture is blurred by large year-to-year swings in Johor, Kuala Lumpur, and Malacca, but overall, there is no upward trend. Apart from the electronics and petroleum sectors, nonmetallic minerals have also received some investment from overseas; however, foreign investors are not enlarging their stakes in Malaysia. They are not pulling out either, but the high levels of FDI that prevailed in the 1990s could be a feature of the past.[11] Only a very detailed examination of the project-level data

---

[11]This possibility may be a source of concern because the contribution of foreign establishments to the total value added of the manufacturing sectors increased from 33.4 percent to 44.2 percent between 1986 and 1999. During the same period, the employment share of foreign subsidiaries also increased from 30.3 percent to 38.1 percent (Tham 2004).

Figure 4.7 Composition of Radio, Television, and Communication FDI Inflows into Malaysia by Source Economy, 1999–2003

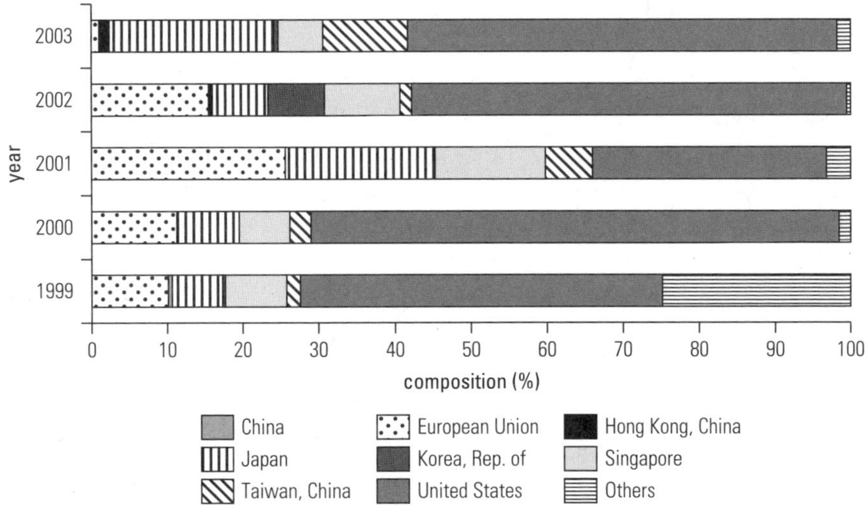

- China
- Japan
- Taiwan, China
- European Union
- Korea, Rep. of
- United States
- Hong Kong, China
- Singapore
- Others

Source: ASEAN FDI Database, 2004.

would show if some niche activities in the major manufacturing subsectors are receiving a disproportionate amount of attention, but such information is not available at present.[12]

Can the sources of FDI yield further insights into the direction in which FDI is nudging Malaysia's industrial evolution? Between 1999 and 2003, investors from the United States, already the biggest source, increased their share, as did investors from Japan and Taiwan, China. The share of Singapore and that of "others" shrank (see figure 4.7). It appears that MNCs from Japan; Taiwan, China; and the United States selectively increased the scale of their operations in Malaysia or at least maintained the level of operations, mainly in electronics. This trend acknowledges the excellence of the manufacturing capabilities that have been created in Malaysia; the relative stability of the workforce; the trustworthiness of local suppliers; and of course, the interests of MNCs, which need multiple sources of supply. Neither the pattern of FDI nor the statements by industry leaders shed more light on which industries might hold the most promise for Malaysia, although there is much talk—but less action—of moving toward a "high-mix, low-volume" production regime. MNCs may still view Malaysia's electronics and telecommunications industries as competitive in terms of prices, quality, and delivery, but the investment data offer few clues as to future deepening of these

---

[12] See Schot and Geels (2007) on the conceptualization of niche development and niche widening.

industries and profitable areas for diversification. Moreover, the geographic redistribution of industry in the region, which is likely as the global economic slowdown of 2008 to 2010 runs its course, increases the uncertainty regarding the future intentions of the MNCs.

Technology Infusion from FDI and Upgrading

A large share of Malaysia's exports of high-technology products (mainly in the electronics and electrical engineering categories) are produced by MNCs or their affiliates (Rasiah 2004). MNCs are also the sources of such exports from China, Thailand, and Singapore (Whalley and Xin 2006). In China, the situation is apparently changing, with more home-grown firms located in the Economic and Technological Development Zones (ETDZs) entering the export market.[13] This change is not happening in Malaysia. Domestic subcontractors producing parts and components for MNCs are not multiplying in Malaysia,[14] and the sparseness of such backward links from MNC operations points to the persistent weakness of domestic entrepreneurship, innovation, and design. This situation reflects the policy of the MNCs to reduce the number of suppliers and to ensure that the ones they contract with can service their needs globally. It also underlines the increasing unwillingness on the part of MNCs to share technology, except in China, where the lure of a large domestic market has forced them to bow to pressure from the authorities to introduce and transfer some cutting-edge technologies and groom local suppliers.

The evidence supporting technology spillovers from FDI is also equivocal.[15] Horizontal spillovers from FDI in manufacturing are weak at best, and Malaysian experience over the past two decades confirms this negative finding

---

[13]There is conflicting evidence on whether Chinese firms are catching up with MNCs. Blonigen and Ma (2007) find that the share of MNC exports from China and their unit values are not rising, suggesting that catch-up is not yet occurring. Wang and Wei (2007) find, however, that the sophistication of exports by Chinese firms, especially those in ETDZs, is on the rise.

[14]Henderson and Phillips (2007) maintain that only Penang developed the institutional capacity (centered on the Penang Development Corporation) to encourage links between MNC production units and local suppliers and to develop a unique program for building skills through the Penang Skills Development Centre. Other states lagged and missed the opportunity. The government's focus on social policies to improve the economic standing of the Bumiputra community may have diluted industrial entrepreneurship, and government efforts to develop state-sponsored heavy and transport industries may have deflected attention from the deepening of the electronics subsector.

[15]There is a large literature on this issue, but the results are mixed. For example, some find overall positive effects (Haskel, Pereira, and Slaughter 2007; Kohpaiboon 2006; Montobbio and Rampa 2005; Yao and Wei 2007), whereas others find no or a negative impact from FDI (Bwalya 2006; Haddad and Harrison 1993; Ng 2006).

(Rasiah 2003).[16] Vertical diffusion of technology to subcontractors of MNCs is more common (Kugler 2006; Noor, Clarke, and Driffield 2002), and such technology transfer is most frequent between MNCs and their joint ventures, affiliates, and subsidiaries. Although a small number of Malaysian firms, such as Pentamaster and Eng Teknologi, have become major suppliers of plant automation equipment and parts by working closely with MNCs such as Dell, Intel, Agilent, and Seagate (all of which have factories in Penang), instances of vertical transfers are few.

Technology diffusion associated with research conducted by MNCs appears to be related to the scale and nature of such research. For example, MNCs such as Intel, IBM, Cisco, and Hewlett Packard have invested in major research facilities in China, Israel, and India.[17] In some cases, such as in India, they have set up specialized research centers in local universities.[18] Episodic evidence suggests that a substantial concentration of research effort, which can extend from some basic research to applied research to technology development, can become the precursor for local start-ups. Gordon Moore, the founder of Intel, writes, "Large firms in high technologies are repositories of unexploited ideas. Restated another way, big firms have natural diseconomies of scope that a cluster of start-ups does not. Therefore, start-ups will prosper in areas where there are also large active high-tech firms" (Moore and Davis 2004: 33). Successful start-ups almost always begin with an idea that has ripened in the research organization of a large company or a university. The experience of Cambridge, United Kingdom; Silicon Valley; and San Diego, California, suggests that in each of these areas, a few axial firms initiated cluster formation assisted by the inputs from world-class universities.[19] Wong (2008) also observes that Singapore has relied on an MNC leveraging strategy that uses a leading global firm to anchor a cluster, expand its role, upgrade its operations, and develop or work with suppliers. MNCs such as Motorola, Intel, Seagate, Agilent, Altera, and Advanced Micro Devices all employ hundreds of engineers in product development, customization, and testing in conjunction with their R&D units in other countries. Firms such as Intel, Altera, and Seagate are also using their Penang-based labs for the development of their latest 28 nanometer chips

---

[16] This finding is also supported by a study of manufacturing enterprises in Indonesia by Blalock and Gertler (2003). Interviews conducted by the authors in the summer of 2008 in Penang also indicated that horizontal spillovers were minimal, and very few individuals left MNCs to start their own firms.

[17] Intel, for example, has an R&D center in Haifa, the first R&D facility for microprocessors outside the United States (Petersburg 2006). Intel's R&D centers in Beijing and Shanghai focus on other research areas, such as replacement BIOS (basic input-output system) and parallel computing, but not processors, mainly because of a lack of qualified personnel in China (Krazit 2007).

[18] A survey of MNCs by Thursby and Thursby (2007) finds that decisions about where to locate R&D activities depend on the size and growth prospects of the market, the quality of local R&D skills, and the protection of intellectual property.

[19] Figure 6.2 in chapter 6 illustrates this concept.

and new read-write heads for hard drives (Tan 2009). However, the nature of the research that is feasible is hampered by the qualifications of the R&D researchers in Malaysia, very few of whom hold postgraduate or doctoral degrees. The small number of patents registered, the limited output of technical papers, and the absence of spinoffs are all indicative of the level of R&D being conducted.

FDI in major R&D activities is determined by the local availability of abundant and high-quality technical workers and of the physical and human infrastructure that supports research. Governments can magnify the attractions of a location by means of fiscal and other incentives. Where skills are sparse, the domestic market is small, and competing locations exist in nearby countries, MNCs are unlikely to expand R&D activities beyond what is needed to promote the effectiveness of local operations. As Rasiah (2007: 24) has pointed out, "In Penang, the supply of R&D engineers and technicians is too small for the [MNCs] to upgrade further into R&D activities."[20] This situation is even truer for Johor. It "looks to remain a platform for the assembly of tail-end activities to support a regional high-tech hub in Singapore" (Rasiah 2007: 18), which has augmented local supplies of technical skills through a liberal immigration policy.[21] Rasiah's (2007: 17) scoring of R&D capital—human and physical—in Malaysia yielded very low means, irrespective of location or ownership exacerbated by "the limits imposed on the import of foreign human capital."

The narrow base of skills, the smallness of the domestic market for high-tech products and services, and the dispersion of the limited R&D capital of the country among three urban centers, none of which enjoys significant agglomeration economies, may explain why Malaysians are unwilling to leave relatively secure jobs in MNCs to start their own firms.[22] These factors may also be the reason industrial clusters in Johor, Kuala Lumpur, and Penang have been slow to expand and to diversify through innovation into new product categories. Given the current stock of technical skills in Malaysia and the modest annual addition of workers with postgraduate- and postdoctoral-level qualifications to this stock, the scope for developing innovative and diversified clusters around a core of electronics industries may exist in Penang and possibly in the Kuala Lumpur area, a point discussed further in chapters 5 and 6.

---

[20] In 2000, there were more than 5,000 scientists and engineers in R&D per 1 million people in Japan. The comparable number for China was 545. In 1998, there were just 160 in Malaysia (Fan and Watanabe 2006).

[21] Starting in the early 1970s with the assembly of consumer electronic products, Singapore moved into computers and peripherals and, more recently, the design and manufacturing of electronic components and wafer fabrication (Wong 2008).

[22] Another possible reason may be that those with working experience at an MNC may leave Malaysia to seek employment abroad. Subsidiaries of U.S. MNCs in Malaysia have a high proportion of local staff in senior management positions. Some of these local senior managers are sent to other subsidiaries worldwide to manage operations there. For instance, when Intel and Motorola established a new factory in China, personnel from Malaysia were sent to manage it. Through their role in managing subsidiaries, some of the Malaysian managers have been absorbed into the senior management of the parent firms (Ariffin and Figueiredo 2004).

5

Leading and Faltering Industries
The Electronics, Auto Parts, and Agro-Processing Sectors

A closer look at three of Malaysia's major manufacturing subsectors can increase our understanding of their innovative capabilities and the degree to which these industries can serve as future leading sectors, not only in Malaysia but also in other Southeast Asian countries at comparable levels of development. These three industries are (a) electronics, (b) auto parts, and (c) palm oil and its byproducts. Together, they generated almost 47 percent of Malaysia's industrial gross domestic product (GDP) in 2005 (table 5.1).

Electronics and Electrical Engineering

Electronics (including electrical engineering) is Malaysia's leading industrial subsector. It was the source of more than two-thirds of exports between 1996 and 2006 (and 37 percent of all jobs in the manufacturing sector). Analysis of product diversification options (see "Opportunities Defined by Mapping Product Spaces" in chapter 3) indicates that most of the top 20 high-density commodities within reach are in the electronics and electrical engineering industry.[1] The effort to

---

[1] For example, new opportunities have opened up at the intersection of the electronics, medical, and information technology–based sectors that could leverage Malaysia's engineering and electronics production and design skills. They could also derive impetus from the demand generated by medical tourism, in particular, for hip and knee replacement and for diagnostic tests, as well as from research and development on aspects of biotechnology in Universiti Sains Malaysia and in some of the government research institutes. There is plenty of scope for advancing lab-on-a-chip technologies (Merkerk and Robinson 2006) and other types of diagnostic devices, as well as body implants. The recent entry of Symmetry Medical, a manufacturer of orthopedic devices from Warsaw, Indiana,

(continued on next page)

Table 5.1 Contribution of the Electronics and Electrical Machinery, Automotive, and Palm Oil Industries to Manufacturing GDP, 2005

Industry	RM million in 1987 prices	Share of total manufacturing GDP (%)
Electronics	3,639	28.0
Electrical machinery	23,043	1.2
Transport equipment	952	12.9
Vegetable, animal oils, and fats	10,667	4.4

Source: Economic Planning Unit.

upgrade the industry by moving up the value chain and creating an innovative cluster has been ongoing for at least a decade and is strongly supported by incentives introduced under the Second and Third Industrial Master Plans.[2] A report by the consulting firm Frost & Sullivan in 2003 (cited in PSDC 2007) identified five new technologies suitable for indigenous development: (a) microelectromechanical systems, (b) photonics, (c) advanced electronic displays, (d) high-density storage, and (e) conductive polymers.

Thus far, the expectations voiced in the mid-1990s have yet to be realized.[3] In 1990, 74 percent of all manufacturing employment was still based on production work, decreasing only slightly to 68 percent by 1998 (Henderson and Phillips 2007). Growth of value added by the electrical machinery subsector that had risen by 46 percent per year during 1971 to 1990 had fallen to just 2 percent (Rasiah 2008) and was paralleled by a steady increase in the share of imports of electronic components and semiconductors (Ernst 2003). The industry's trade balance, which improved through the 1990s, turned negative after 2000 (Rasiah 2008). The economywide growth of total factor productivity (TFP) in electronics and electrical engineering slowed from 14.1 percent per year during 1990 to 1995 to 2 percent per year from 1995 to 1999 (Ernst 2003). In Penang, the growth of TFP in electronics and electrical engineering declined from 8.9 percent per year during 1990 to 1995 to −0.5 percent per year between 1995 and 1999 (State Government of Penang 2001).

(*continued from previous page*)
into Penang's embryonic biomedical sector is a welcome development. The firm was attracted by the low cost of wages and other expenses, the local precision-machining skills vital for making cobalt-chromium steel hip joints, and the ease of marketing to and serving regional hospitals conducting joint replacement surgeries.

[2] On cluster development and innovation in clusters, see Yusuf, Nabeshima, and Yamashita (2008).

[3] The four pioneering East Asian Tigers also started out with the simple assembly of electronic products and, over time, assimilated the technology to move up the value chain and raise domestic content. The Republic of Korea relied less on foreign direct investment than did the other three (Hobday 1994).

Few Malaysian firms have entered the product niches identified by Frost & Sullivan, as is apparent from five developments. First, in microelectromechanical systems, AKN Technology is the only Malaysian company now operating, and neither its entry into the subsector nor its modest success in the export market is linked to any specific government incentives (PSDC 2007). The government research institute MIMOS (Malaysian Institute of Microelectronic Systems) and a couple of the leading local universities have made some progress in initiating microelectromechanical systems research and grid computing, but specific channels through which links with the industry can be consummated have not materialized. Thus far, MIMOS has not fulfilled the catalytic role envisaged by the government somewhat akin to that of the Industrial Technology Research Institute in Taiwan, China. Second, Finisar, in Ipoh, is the only prominent player in the fiber-optic components industry. Most photonics companies in Penang are LED (light-emitting diode) based, with a few local companies, such as Optometrix, beginning to go beyond components manufacturing and acquiring competencies in designing, fabricating, and building test systems. Spinoff benefits of LED applications have accrued mainly to multinational corporations (MNCs), with little by way of participation by Malaysian companies. Third, no progress has been made in the areas of technology related to advanced electronic displays.[4]

Fourth, high-density storage (both magnetic and optical)—that is, the disk drive industry—is probably the only industry that has demonstrated "technological advancement" in Malaysia. A leading Malaysian firm, Eng Teknologi, produces actuators, stators, baseplates, and spindle motor components for hard disk drives plus other items. Penang-based Eng Teknologi has emerged as one of the top five component suppliers for hard disk drives, among other components ("Corporate: EngTek" 2007). Actuators account for 25 percent of the firm's sales. Other services include precision engineering and tooling and components machining (Tan 2007). Eng Teknologi has also become an MNC, with facilities established in Dongguan, China (in 1996); the Philippines (in 1997); and Thailand (in 1998) (Hiratsuka 2006). Actuators produced at the Dongguan facility are mainly for 3.5-inch disks for desktop, external hard drives, video game consoles, and digital video recorders (Tan 2007). Its facility in Thailand produces actuators for 2.5-inch disks, mainly for mobile devices, whereas the facility in the Philippines produces actuators for 3.5-inch hard disks, mainly for servers (Tan 2007). All of Eng Teknologi's expansion seems to be occurring outside of Penang. The Dongguan facility increased its output from 1 million actuators per month to 2 million in the first quarter of 2008, and to 3 million by the end of 2008 through an investment of RM 15 million (Tan 2007). In addition, the development of local suppliers in China is threatening the traditional suppliers based in Penang. Although about 80 percent of the inputs used in actuator production in the past came from other

---

[4]Taiwan, China, is the global leader in advanced electronic display technology (PSDC 2007).

Penang-based firms, Eng Teknologi currently sources about 50 percent from Chinese suppliers. With the rapid development of suppliers in China, Eng Teknologi may source the majority of its inputs from China, and the share of domestic components (mainly from Penang) may become as small as 20 percent (Dwyer 2007). Furthermore, Eng Teknologi is establishing two research and development (R&D) centers: one in Senai, Malaysia, and the other in Dongguan, China. The center in Senai will concentrate on technology development for baseplates. The center in Dongguan will concentrate on product development, especially that of actuators. Although wage costs in Malaysia and Thailand are higher, firms continue to concentrate on labor-intensive operations. Moreover, these operations are now threatened by lower-cost producers in the Pearl River Delta and Vietnam, whereas at the other end of the spectrum, high-value-added developmental and testing activities are increasingly concentrating in Beijing and Shanghai and in Singapore.

The fifth and final development, that of conductive polymers, did not take off either (PSDC 2007). The conductive polymer sector has remained a relatively unsupported, low-profile industry, with the local companies engaging in "blending" services, while the advanced plastics (epoxies and adhesives) are imported from Japan and the United States.

If the trends in productivity and value added are to be improved and technological capacity enhanced, Malaysian firms will need stronger support from local tertiary research institutions and overseas providers of technological support to upgrade research, development, and design capabilities. This collaboration would focus on specific niches in line with Malaysia's technology development objectives (for example, chip design, embedded software, and photonics). Malaysian universities need to introduce courses and postdoctoral programs that are customized to the specific needs of Malaysia's electronics firms. More than 60 percent of the 167 electric-electronics firms surveyed in Penang and Johor in 2005 reported that university curricula and lab equipment in Malaysia do not prepare students for the kinds of jobs available, and firms must invest in longer training periods of six months or more for new recruits (Rasiah 2008). Nine flagship electronics firms considered the private Multimedia University in Kuala Lumpur as the most active in soliciting suggestions from companies and tailoring its engineering degree curriculum to meet the needs of industry. For example, Multimedia University and also Universiti Teknologi Malaysia have worked hand in hand with Altera, located in Penang, to customize programs and a curriculum to improve the practical skills of graduates (Rasiah 2008). Altera has been especially generous in sharing its training practices with the two universities and in transferring some equipment to them.

One outstanding success story associated with the development of the electronics industry is the Penang Skills Development Centre (PSDC), which has become a world-class training center. But similar organizations have been slow to emerge in other parts of the country (Henderson and Phillips 2007). PSDC's partnership with the firms in Penang to design and conduct training enhanced the relevance of

PSDC's programs. The Selangor Human Resource Development Centre, modeled after PSDC, has made significant strides recently, but others, such as the Johor Skills Development Centre, still lack sufficient capabilities (Rasiah 2008).

Improvements in the overall quality of training and development and the creation of shared research laboratories are among the developments that are needed to ensure the longer-term growth of Malaysia's industries, according to a new report by the Industry Research Task Force of Penang (PSDC 2007). The report emphasized the importance of industry clusters composed of innovative small and medium-size enterprises (SMEs) as essential for moving up the electronics and electrical engineering value chain. SMEs need to adopt productivity- and technology-enhancing tools introduced by the information and communication technology (ICT) industry to connect their business processes to global customers and become part of the value chain. Although finance cannot initiate innovation, easier access to grants by simplifying application and approval processes under the guidance of the Small and Medium Industry Development Corporation would surely ease one widely cited constraint. Finally, better access to market research and intelligence on new products and technologies would be facilitated by a repository of information on technological advances or a resource center to assist firms in locating relevant solutions for their needs.[5]

Auto Parts Industry

The Malaysian auto industry can be categorized as a case of failed upgrading despite generous incentive policies.[6] Hopes ran high after Proton Holdings Berhad, the first national auto company, began assembling cars in July 1985 using technology and parts sourced from Mitsubishi. But despite years of infant industry protection,[7]

---

[5]PSDC (2007) identifies the types of firms in Penang that need to further strengthen their competencies as manufacturers of (a) semiconductor and printed circuit board assembly process equipment, (b) semiconductor test systems and manufacturing consumables, (c) wafer fabrication equipment, (d) advanced substrate technologies, (e) advanced polymers, (f) sensor and sensing devices, (g) disk drive platters, (h) LED applications and solar cell technologies, and (i) wireless devices, as well as firms involved in (j) industrial, mechanical, and electronic outsourcing with rapid prototyping; (k) development of technical and engineering software and ICT; and (l) design and manufacture of radio frequency–microwave, portable medical electronics and medical mechanics.

[6]Malaysia seems to have fared poorly on all of the following dimensions of upgrading: (a) production at levels of price, quality, and delivery that permit significant growth of exports; (b) manufacturing of products that embody increasing value added; and (c) growing levels of local (indigenous) inputs or links.

[7]Import duties ranged from 40 to 300 percent to protect the domestic market ("Proton Bomb" 2004). Further technology transfers from Mitsubishi and engine, transaxle technology, and design capacity acquired through the takeover of Lotus in 1996 have also not enabled Proton to overcome the hurdles it faces.

Proton has struggled to achieve viability. Its share of the domestic market for sedans, which rose to 57 percent in 1993, had fallen to 32 percent in 2006. In 2007, this share dropped to 24 percent, with the carmaker reporting a net loss of US$169 million and looking for a strategic partner to provide it with the capital, technology, and global market reach it urgently needed. The domestic market share for sedans rose to 26 percent with the introduction of new models in 2008 (by which time Proton had produced a total of 3 million vehicles), but the underlying weaknesses remain ("Malaysian Carmaker" 2008). Proton's travails resemble those of other smaller carmakers, such as Rover, Jaguar, Volvo, Land Rover, and Saab; all of these carmakers ended up as subsidiaries of global giants, but despite considerable efforts by their parent companies, they have struggled to achieve profitability. The Chinese manufacturer SAIC (Shanghai Automotive Industry Corporation) took over the Rover brand. Ford sold the Land Rover and Jaguar brands to Tata Motors in 2008, and both Saab and Volvo were up for sale in 2009 as the American auto giants attempted to cut mounting losses. The inability of Proton to evolve into a major multinational firm such as Hyundai after two decades of trying has affected the fortunes of the auto components industry and has hampered parts suppliers from participating in the rapid growth of the trade in global automotive components.

The backdrop to the weak performance of Malaysia's automobile industry is the contrasting relatively favorable global and regional (Southeast Asian) environment throughout the period until the middle of 2008. Growth of trade in automotive products was 50 percent higher than that of office and telecommunications equipment (the largest component of global manufacturing trade in 2006), during 2000 to 2006. Moreover, exports of auto products from Asia were averaging a 12 percent annual growth from 2000 to 2006 (World Trade Organization 2007).

In 2006, automotive products constituted only 0.6 percent of Malaysia's merchandise exports, in contrast to 1.7 percent in Indonesia; 1.8 percent in Taiwan, China; 3.2 percent in the Philippines; and 7.6 percent in Thailand (Ravenhill and Doner 2007). In value terms, Malaysia's automotive exports in 2006 were less than one-tenth of Thailand's and only slightly over half of the exports of Indonesia and the Philippines, countries with substantially weaker national technological capabilities.

Malaysia's imports of automotive products, on the other hand, constituted 2.5 percent of total merchandise imports in 2006, with a value that was three times that of exports. A fourfold increase in the imports of components during 1996 to 2006 was not matched by the twofold rise in domestic production. By 2005, imports constituted 47 percent of the domestic consumption of automotive components. This surge of imports points to Malaysia's lack of a competitive advantage in this industry. As trade regimes have changed (both through the World Trade Organization and the Association of Southeast Asian Nations), this lack has resulted in massive inflows of automotive imports from neighboring countries, such as Thailand, which displaced the European Union (EU). Overall, export growth of the automotive sector during 1996 to 2005 averaged 12.0 percent annually, slightly faster than the 11.2 percent annual growth rate for all manufactured exports.

The share of unsophisticated parts in Malaysia's auto exports was almost unchanged from 1988 to 2006. Exports of high-tech and higher-value parts, albeit exhibiting a positive annual growth rate, still constituted less than 10 percent of overall auto components exports in 2006. In 1988, Malaysia started out with the lowest share of low-tech products in its auto components exports, but by 2006, these products constituted a higher share in total components exports than was the case for either Indonesia or Thailand. Relatively unsophisticated activities, such as stamped-metal parts and molded-plastic parts, dominated Malaysia's auto parts exports, among which the most notable items were lighting equipment, batteries and battery parts, safety glass, springs, oil and gas filters, wire harnesses, brake parts, and bumpers.

Productivity growth during 1996 to 2005 of sales value per employee increased at an annual average rate of 2.3 percent, which was the lowest for all the sectors listed in the Third Industrial Master Plan. The average annual growth rate in productivity of the manufacturing sector as a whole was 8.5 percent for the same period.

More than 90 percent of the total 769 firms in the transport equipment sector were SMEs (Ravenhill and Doner 2007). Their small size and sparse links with the auto assemblers and with other suppliers limited their capacity for R&D, implying that they depended on the design technology provided by the automobile manufacturers. Moreover, most of the first-tier suppliers in the country, which had the potential for R&D, were foreign owned, but the small size of the Malaysian market (for example, until June 2003, foreign investors were limited at best to a 51 percent share of projects in the auto industry) discouraged even those firms from conducting product development locally.

Component producers received more favorable treatment than did lower-tier suppliers serving foreign assemblers in other Southeast Asian economies. In particular, high levels of protection and restrictions on the share of imported completely built-up units to no more than 10 percent of the domestic market have helped to promote the local auto parts industry but have failed to create competitive auto parts suppliers. Companies undertaking design, R&D, and production of qualifying automotive component modules or systems were eligible for pioneer status. This designation allowed firms to claim a tax exemption of 70 percent of the statutory income for five years[8] and an investment tax allowance of 60 to 100 percent on qualifying capital expenditure.

The problem with the development of an automotive industry in Malaysia is similar to that in Indonesia, where the government tried to develop the assembly operation without first nurturing the necessary supply base domestically (Aswicahyono, Basri, and Hill 2000).[9] The Malaysian government tried to

---

[8]The tax exemption is 100 percent for high-tech activities, for SMEs, and for firms in the Eastern Economic Corridor.
[9]The development of the aircraft industry in Indonesia shared the same problem (Aswicahyono, Basri, and Hill 2000).

correct this deficiency. However, the very policies that seek to promote the local automotive parts industry have proved to be its nemesis. By limiting the access of foreign assemblers to the domestic market and by imposing restrictions on foreign investment,[10] the government largely ruled out the possibility of Malaysia becoming a regional hub for foreign auto assemblers and for foreign first-tier components suppliers, who largely turned to Thailand. The transport equipment sector attracted only 6.2 percent of all approved investments in manufacturing during 1996 to 2005, of which foreign investors contributed 41 percent, compared with their 56 percent share in total investment (Ravenhill and Doner 2007). A guaranteed domestic market reduced incentives for auto component firms to become more efficient, to upgrade their skills, and to consolidate their operations into larger, more viable operations that could realize economies of scale.[11] Proton interfered with the process of consolidation by decentralizing the procurement of components, with the intention of involving vendors more closely in platform design. Unfortunately, this policy resulted in a proliferation of vendors when different companies were appointed to supply the same component for different models, an outcome that Proton is trying to reverse. The lack of opportunities to produce in volume for the domestic market, along with second-round repercussions caused by the lack of investment funds (some government SME banks were reluctant to lend because of the perceived problems and lack of profitability in the industry), has created a vicious cycle.[12] The acquisition of Lotus was supposed to reduce or eliminate the dependence on foreign engine technology. Unfortunately, Lotus lacked the expertise in camshaft design to produce the needed engines; consequently, Proton continues to rely on Japanese auto companies for the requisite technology.

The future of the auto components industry does not look promising, given its inability to upgrade and to increase export penetration, despite the prevailing government incentives. Furthermore, the product space analysis in chapter 3 suggests that the opportunities for Malaysia to specialize and diversify into new automotive products and components are low, compared with those related to other

---

[10] The restrictions were loosened in 2003 to allow majority foreign ownership but not 100 percent foreign ownership.

[11] Proton has been exporting a small number of cars to Australia, South Africa, and the United Kingdom and is now aggressively seeking to expand export markets in Thailand through collaboration with the PNA Group of Indonesia that is aimed at the taxi business. It is also seeking to expand in China, through collaboration with Young Man Automobile Group, and in other emerging markets. The United Kingdom has been the most successful destination, with the sale of 1,518 cars in 2008.

[12] Consequently, the companies that were excluded from the Vendor Development Program, along with those that made an early decision not to rely primarily on Proton for their sales, prospered. They sought joint ventures or technological tie-ups with foreign—especially Japanese—partners. Most of these firms' operations were export oriented: the Malaysian Automotive Component Parts Manufacturers estimated that between 50 and 60 of its 103 members engaged in exports (Ravenhill and Doner 2007).

high-value commodities, such as electronics and electricals and engineering goods. Figure 3.20 (in chapter 3) shows that electronics and electrical products are mostly in the densest region of Malaysia's "open forest," whereas appendix table B.6 shows that only one commodity at the four-digit Standard Industrial Trade Classification (revision 2) level exhibits a high probability of potential diversification opportunities. The other automobile industry products are ranked 183, 220, 256, 323, and 363 in the highest-density upscale products.[13]

Palm Oil, Biodiesel, and Food Products

During the 1950s, Malaysia relied on rubber as a main agro-based industrial output. The government invested in R&D on rubber technology to improve the yield, and Malaysian firms such as Top Glove and Supermax have become the leading global exporters of latex gloves, thread, and catheters, as well as of conveyor belts, floor coverings, and rubber footwear. At the same time, the government actively sought diversification from rubber and decided to expand the palm oil sector starting in the 1960s. The oil palm tree is native to West Africa, and that region was the source of the world's palm oil for lubricating purposes and for soap throughout the 19th century.[14] The oil palm was first introduced into Malaysia as an ornamental plant in the 1879s. Later, in 1917, William Sime and Henry Darby established a 500-acre plantation at Tennamaran in Selangor.[15] To reduce rural poverty and with the help of a newly created agency, the Federal Land Development Authority, Malaysia rapidly expanded the acreage devoted to palm oil in the late 1960s and 1970s. By 1970, the output of palm oil more than quadrupled to 400,000 tons, compared with 90,000 tons in 1960 (Yusof and Bhattasali 2008). Further development of upstream and processing operations promoted through the first and second industrial master plans (of 1986 and 1996, respectively) enabled Malaysia to emerge as the largest exporter of palm oil in the world and the second-largest producer of palm oil in the world.[16] In 2008, Malaysian production of palm oil had reached 17.3 million tons, and it accounted for 52 percent of the global oils and fats exports (Sumathi, Chai, and Mohamed 2008).[17]

---

[13]These products are, respectively, passenger cars, chassis fitted with engines for cars, other car parts and accessories, road tractors and semitrailers, and bodies for motor vehicles.
[14]Archaeological evidence from the Arab Republic of Egypt shows that palm oil was in use 5,000 years ago (bin Abdul Hamid 2009).
[15]Sime Darby grew into one of the largest agro-producing companies in Malaysia until it was acquired by Malaysian interests in the late 1970s through a famous "dawn raid" on the London Stock Exchange.
[16]Indonesia is the leading producer of palm oil ("Indonesia/Malaysia: Palm Oil" 2008).
[17]In 2007, 3.9 million hectares were devoted to oil palm production. Output consists of crude palm oil, palm kernel, crude palm kernel oil, and palm kernel cake (MPOB 2007). Residues are used to fertilize the soil and for biogas production. Total exports of oil palm products in 2006 were 20.2 million tons.

In recent years, Malaysia has begun diversifying into biofuels.[18] By 2007, Malaysia had an installed capacity of 1 million metric tons, almost double that of the next country, Indonesia, with 7.6 billion liters (Malaysian Palm Oil Council 2008; Sharma and Singh 2009).[19]

The oil palm is a better alternative than other crops for producing biodiesel for three reasons (Sumathi, Chai, and Mohamed 2008). First, in terms of productivity (and hence profitability), it is the highest-yielding crop, with an average of 3.7 metric tons per hectare, about five times higher than rapeseed and seven times higher than soy. The oil palm is also the most productive oilseed in the world. "A single hectare of oil palm may yield . . . 6,000 liters of crude oil according to data from Journey to Forever. For comparison, soybeans and corn—crops often heralded as top biofuel sources—generate only 446 and 172 liters per hectare, respectively" (Butler 2006). "By Curran's calculations,[20] some firms in West Kalimantan are seeing a 26 percent annual internal rate of return over a 25-year period, an astounding number" (Butler 2007). Because of this high yield, the cost of producing 1 ton of oil from palm is cheaper at US$228 than that of producing oil from other crops such as soybean (US$400) and rapeseed (US$648) (Lam and others 2009).[21] Second, the more immediate environmental benefit is that of energy efficiency. In particular, the ratio of energy output to input (energy used to produce 1 ton of oil) is almost three times higher for palm oil than for soybean and rapeseed oil. Jaafar, Kheng, and Kamaruddin (2003) estimated the renewable energy potential for oil palm biomass at more than RM 6 billion (annually), second only to forest residues, which had an energy value of almost RM 12 billion. Third, biodiesel is relatively benign with regard to its implications for the environment and for global warming. It reduces greenhouse gas emissions to the tune of 50 to 60 percent relative to gasoline, but the aggregate environmental costs depend on whether carbon-rich ecosystems are destroyed to make room for oil palm plantations (Scharlemann and Laurance 2008; World Bank 2008e).[22]

Global prospects for biodiesel as a source of renewable energy would be promising if fossil fuel prices return to the levels of the first half of 2008. According to estimates made before the global recession that took hold in the latter half

---

[18] The first biofuel was developed by Rudolf Diesel, the inventor of the diesel engine, in 1900. His biofuel was based on peanut oil (Sharma and Singh 2009).

[19] In 2007, however, Malaysia produced only 80,000 tons because of the high price of palm oil ("Bugs in the Tank" 2008).

[20] Dr. Lisa Curran is a biologist who has spent more than 20 years in Borneo. In a series of papers, she has documented the emergence of oil-palm plantations on the island.

[21] The palm oil industry also does not depend on government subsidies, as may be the case in other countries for producing crops for biodiesel (Lam and others 2009).

[22] However, any biodiesel created from food crops faces the criticism that its diversion of food results in higher food prices and increases undernourishment and malnourishment. Moreover, there are concerns for biodiversity, such as protecting the habitat of orangutans on the islands of Borneo and Sumatra (Lam and others 2009).

of 2008, "Total world consumption of marketed energy is projected to increase from 447 quadrillion Btu in 2004 to 559 quadrillion Btu in 2015 and then to 702 quadrillion Btu in 2030—a 57 percent increase over the projection period" (EIA 2007: 5). Transportation is expected to account for about half of the total projected increase in global oil use between 2003 and 2030. Transportation consumed 30 percent of this global energy production (compared with close to 40 percent for power generation), 99 percent of which was supplied by petroleum (EIA 2007). In 2006, liquid biofuels (mainly ethanol and biodiesel) accounted for just over 1 percent of global renewable energy (16 million tons of oil equivalent out of 1,430 million tons of oil equivalent) and less than 1 percent of global crude oil supply of 4,800 billion liters (IEA 2006). However, if the plans to increase the production of biofuels come to fruition, there could be a fivefold increase in the share of biofuels in global transport—from just over 1 percent today to around 6 percent by 2020 (IEA 2004; Rothkopf 2007). Forecasts from "Clean Energy Trends 2007" reveal that global markets for biofuels (manufacturing and wholesale pricing of ethanol and biodiesel) reached US$20.5 billion in 2006 and are projected to grow to US$80.9 billion by 2016, assuming that global growth rates return to past trends by 2010 to 2011 and petroleum prices continue trending upward from levels reached in mid-2009 (Mankower, Pernick, and Wilder 2007).[23]

Malaysia's own interest in exploiting biodiesel as an alternative source of energy has strengthened since the beginning of this millennium. Malaysia's remaining oil reserves are expected to last for another 20 years and its natural gas reserves are expected to last 50 to 60 years (Basiron, Balu, and Chandramohan 2004).[24]

Palm oil emerged as a diesel substitute in Malaysia during the mid-1990s (Choo and others 2005). In March 2006, the Malaysian prime minister launched Envodiesel, which blends 5 percent processed palm oil (vegetable oil) with 95 percent petrodiesel (MIDA 2006). This program is designed to enhance the economic welfare of the country's small-scale plantation holders (Mathews 2007). By July 2006, licenses had been issued for 52 plants capable of producing 5 million tons of biodiesel annually.[25] According to Najib Razak, then deputy prime minister (he succeeded Abdullah Badawi as prime minister on April 3, 2009),

---

[23]Biofuel development across the world is being driven as much by politics as it is by economics. Its future will hang on the policies of hydrocarbon fuel; on the total costs of producing and distributing biofuels; on technologies for producing biofuels from a variety of feedstocks that have a small carbon footprint (for example, switchgrass and algae) and do not compete for land with other valuable food crops; and on automotive technologies. The biofuel boom of 2006 to 2008, now in 2009 seems to be heading to at least a temporary biofuel bust following the decline of petroleum prices in 2008 and the near-term prognosis of slow growth (Carey 2009).

[24]In 2008, petroleum production in Malaysia averaged 600,000 barrels per day, and Malaysia should remain a net exporter of oil very likely through 2011.

[25]However, as of the end of 2007, only seven to eight projects had actually been initiated.

"It would be a mistake if Malaysia, as a major palm oil producer, did not tap the huge potential in the biodiesel market particularly in meeting the demand in Europe and the United States" (Mathews 2007).[26] Malaysian researchers are focusing on palm oil methyl ester to satisfy the EU's blends (diesel engine modes of transportation in the EU have been geared to run on 5 percent methyl ester blend) and the European Norm 14214 standard. To eventually produce 5.2 million tons of biodiesel per year, Malaysia would need to set aside 6 million tons per year of the oil palm as feedstock, nearly 40 percent of its crude palm oil production (Basiron 2005).[27]

There is a downside to increasing the acreage under oil palms. Beyond the loss of forest ecosystems and associated biodiversity, the release of greenhouse gases from the exposure of or removal of deep peat soil, and adverse social effects on the local communities dependent on services and products provided by these forest ecosystems, the production of palm oil, as currently practiced, can be quite damaging to the environment. Production of crude palm oil generates tons of solid oil wastes, palm fiber, and shells, not to mention palm oil mill effluents, which are a polluted mix of crushed shells, water, and fat residues that have been shown to severely degrade aquatic ecosystems. Furthermore, the liberal use of petroleum-based pesticides, herbicides, and fertilizers ensures that most palm oil cultivation not only pollutes on a local level but also contributes to nitrogen oxide greenhouse gas emissions. Considering that Malaysia is held to be one of the most efficient producers, production in other parts of the world may be even more environmentally destructive. Indonesian oil palm plantations are so damaging that after a 25-year harvest, oil palm lands are often abandoned for scrubland. Major Malaysian producers such as Sime Darby have cut down on pollution by adopting a closed loop production regime, which recycles or uses all byproducts. Moreover, these producers use only shallow category 1 peat soils. There is currently a cap on the acreage allocated to oil palm in Malaysia (bin Abdul Hamid 2009).

Biofuels production is being promoted in most developed nations using protective tariffs and direct production (and consumption) subsidies to this industry (developing countries such as Brazil also fall in this category). From the perspective of future industrial development in Malaysia, the question that needs to be addressed is whether the expansion of the biofuel industry will be economically viable using current technologies, after discounting the environmental costs. Hence, the challenge for Malaysia is to avoid "costly" incentives such as subsidies, to implement well-defined regulations, and to realistically account for the actual

---

[26] The European Union has proposed that 5.75 percent of all fuel used in transport should come from biofuels by the end of 2008, rising to 10 percent by 2020. The United States aims to replace 15 percent of fuels for transport with biofuels by 2022, rising to 30 percent by 2030 ("International: Land Conversion" 2008).

[27] Research on the oil palm and on palm oil derivatives is conducted by the Malaysian Palm Oil Board and major companies.

benefits (environmental and economic, as well as social) from increased biodiesel production. There may well be a need to invest substantial private as well as public resources in R&D; to develop small-scale biodiesel production; and to improve the existing technology for using biomass to produce biofuels.

Although palm oil accounted for over one-third of agricultural value added in 2005, the potential of other food crops should not be neglected. Recent trends suggest that the international demand—and prices—of many food products will likely remain high, offering opportunities for Malaysian exports of fresh and processed farm products. Among the most promising are cocoa, pepper, vegetables, fruits, flowers, and herbs. Together they make up about one-fifth of the agricultural GDP and about 10 percent of exports (in 2005). Agriculture has the makings of a growth industry; however, exploiting the possibilities for increased value added through processing and packaging and for diversification calls for investment in research (in genetically modified technology and cultivation techniques, for example) and advances in food processing technologies. The food processing industry is a major subsector—the largest single industry in the EU, for example—and it is often neglected because it lacks glamour.[28] However, given Malaysia's comparative advantage in resource-based products, the improving market prospects for foodstuffs, and the environmental constraints affecting many countries, food processing is an industry to develop alongside oil palm and fisheries.

---

[28] It is also low on the scale of research intensity and is environmentally unfriendly on at least two counts: the release of pollutants and the heavy use of water.

6

Can Southeast Asian Tiger Economies Become Innovative?

Industrial Location

The trends in the composition of trade and distribution of investment among industries provide a perspective on an economy's evolution and potential. These two factors indicate where production and technological capabilities are accumulating, how much deepening and diversification are occurring, and what the scope is for gains in productivity through embodied technological change and learning by doing. Another equally vital perspective is derived from the distribution of the production activities and investment among the main urban centers. Research on urban economies indicates that productivity and innovation are linked to the size and other characteristics of urban centers. In other words, factor productivity and the likelihood of business innovation, whether in electronics or in finance, are affected by location. Building innovation capability starts with location.

There are four findings of relevance for Malaysia as its business community explores new industrial directions.[1] First is the evidence of substantial agglomeration economies as city size grows.[2] These agglomeration economies derive from a number of sources including (a) the rising scale of the local product market; (b) the size and density of the local market for factors (especially labor skills);

---

[1] This section is based on Yusuf (2007).

[2] See Duranton (2008) for a framework analyzing urbanization and the growth of cities in developing countries. See Au and Henderson (2006a); Coulibaly, Deichmann, and Lall (2007); and Overman, Rice, and Venables (2007) for various ways by which to measure urbanization and localization economies. See Rosenthal and Strange (forthcoming) for the link between agglomeration and entrepreneurship and Kahn (forthcoming) on the cost of agglomeration. See Spence, Annez, and Buckley (2009) for a recent survey on the relationship between urbanization and economic growth.

(c) the concentration of many different industries, business services, and research and development (R&D)–related services, which can promote fruitful trading and subcontracting (such economies are usually denoted as urbanization economies as distinct from localization economies, which arise from industrial specialization);[3] and (d) the heterogeneity of talents that can help to spark fresh ideas and innovation. Although it is difficult on the basis of the available econometric evidence to determine the optimal economic size of a city,[4] there are grounds for claiming that a metropolis that has fewer than 5 million or 6 million people is unlikely to realize the full benefits of scale and agglomeration economies.[5] In fact, as Overman and Venables (2005: 27) observe, "the available evidence suggests that, at least from an economic perspective, being oversized is much less costly than being undersized."[6] The mega-urban region presents broader market opportunities—and "a long tail" of promising niches[7]—and is more hospitable to the emergence of new clusters.[8]

---

[3] See Yusuf and others (2003). Although competition is generally viewed as having a positive effect on productivity, firms may choose to avoid agglomeration and clusters of similar producers to minimize competitive pressures (Belleflamme, Picard, and Thisse 2000).

[4] The optimal economic size is the point at which the marginal social benefits are maximized.

[5] Reviewing the evidence on agglomeration economies, Rosenthal and Strange (2004) find that with each doubling of city size, the urban gross domestic product can increase by between 3 and 14 percent. Venables and Rice (2005) estimate that a doubling of the population of a city can raise productivity by 3.5 percent. J. Vernon Henderson concludes, from his assessment of the size and productivity of cities in China, that most are suboptimal in size—that is, below 5 million (Au and Henderson 2006b)—and that productivity gains from an expansion would be about 4.1 percent if the city were 20 percent below optimal, but as much as 35 percent if the city were half the optimal size (Henderson 2004; Rosenthal and Strange 2004). Although evidence on urbanization economies (arising from industrial diversity) is mixed (and greater for some industries than others), that for diversity of high-tech industries is much clearer. Overman and Venables (2005: 18) find that a 1 standard deviation increase in the diversity index raises productivity by 60 percent. See also Deichmann and others (2005). A review of the theoretical literature on city size and growth can be found in Henderson (2005).

[6] A summary of the literature on localization economies is contained in Overman and Venables (2005). A study for Korea suggests that "a 1 percent increase in own industry employment increases productivity by 0.08 percent; i.e., moving from a city with employment of 1,000 to a city with 10,000 would raise productivity by over 70 percent" (Overman and Venables 2005: 18).

[7] See Anderson (2006) on the opportunities presented by the "long tail."

[8] Co-location of component and module suppliers and assemblers, which permits just-in-time delivery, reduces time costs, and allows the advantages that accrue from collaboration, research, and design of products, favors proximity among producers and suppliers—factors that further strengthen agglomeration effects (Harrigan and Venables 2006; Hillberry and Hummels 2005). Knowledge-intensive industries are becoming more

(continued on next page)

Younger industries that are buoyed by emerging technologies and are dependent on knowledge spillovers from research centers, for example, are most likely to locate in large urban agglomerations (Desmet and Rossi-Hansberg 2007).

Second, urban scale is correlated with greater innovation, as proxied by patenting and by the publication of scientific papers. It is also conducive to the cocreation of new products or services by firms in conjunction with consumers, other firms, and research entities.[9] Such cocreation is plausible because strategically located, large urban centers are also the centers of higher education and of research by corporate as well as public entities. Where such urban centers are able to create an open learning environment that is conducive to ideas trafficked through interpersonal exchanges and to experimentation, innovation is more likely to be "superlinear" and can be thought of as arising from the economies of openness and heterogeneity.

The presence of research universities and research institutes, either public or affiliated with major local firms or multinational corporations (MNCs), is a third factor supporting innovation in cities and aiding the cumulative growth of knowledge that underpins scientific advance (see Furman and Stern 2006).[10] These research facilities not only are a source of much-needed technical skills and knowledge spillovers, but are also emerging as essential sources of consulting services, patents, tacit knowledge, and entrepreneurship, which are critical to the success of basic science– and information technology–based industries (Yusuf and Nabeshima 2007).[11] The presence of universities also makes it easier for firms to recruit staff members. This factor encourages firms to establish production facilities and offices in cities that are home to reputable tertiary institutions. The better universities that promote research are most often located in the major European and North American cities; in East and South Asia, they are overwhelmingly found in the capital cities or the very largest urban centers. The ability of these regions to retain talent by providing attractive jobs, amenities, social services, and housing is, as Richard Florida (2002, 2005a, 2005b) has noted, a necessary

---

(*continued from previous page*)
concentrated in areas where there is an abundance of skilled and technical workers (Midelfart-Knarvik and others 2000). Moretti (2002) finds that most of the spillovers are from firms that are higher on the scale of technology. On the necessary conditions for growing industrial clusters, see Yusuf (2008).

[9] The scope and potential for cocreation are discussed by Prahalad and Krishnan (2008).

[10] In the United States, the major federal laboratories are better able to conduct interdisciplinary research and can afford expensive equipment and facilities. Universities have the advantage of a steady flow of students, which affects the nature and culture of research (Bozeman 2000).

[11] A rich literature is accumulating on this topic. Bettencourt, Lobo, and Strumsky (2007) find evidence of superlinearity in patenting among metro regions in the United States; that is, larger regions generate more patents. See also Bettencourt and others (2007) and Carlino, Chatterjee, and Hunt (2007).

condition for their economic success.[12] In fact, by contributing to the cultural life of cities and creating a special ambience, universities can help to make cities "stickier" both for talent and for industry.

A fourth factor favoring bigger cities is that they are likely to house larger stocks of entrepreneurial capital and to encourage a more active exchange of ideas and trading of intellectual property. Moreover, entrepreneurs in large cities are often plugged into constellations of supporting networks that can assist with new starts (Gans and Stern 2003; Glaeser 2007, 2009; Nijkamp 2003). In the case of Australia, for example, the existence of a number of medium-size, widely separated cities has constrained urbanization economies, networking, and innovation.

The bane of the megacity is congestion and the associated higher land and transport costs, increased environmental pollution, and higher living expenses. Other problems can arise (a) from the lack of interjurisdictional coordination to develop, maintain, and regulate infrastructure as well as other services; (b) from weaknesses of governance; (c) from difficulties in attaining the desired level of fiscal effort and in supplementing this effort through funds from capital markets; and (d) from the often daunting challenges of managing the finances of a metropolis.[13] These last three are not only related to the size of cities. Furthermore, research has uncovered no clear relationships between size and environmental pollution.[14] However, rental costs do tend to be higher in the mega city, which affects wages and, hence, induces some low value-adding and land-intensive activities to migrate to the suburbs or to smaller cities (see Overman and Venables 2005).

---

[12]The authors' own recent research (Yusuf and Nabeshima 2007) and that of Florida and his collaborators (Florida and others 2006) have made clearer the seminal role that universities are likely to play in the knowledge economy of the future (Etzkowitz 2002; O'Mara 2005). The university is becoming the axis of a creative hub, not only in generating the knowledge that underpins innovation but also in mobilizing talent, testing ideas, and creating a tolerant social climate. Florida's study of Sweden's regions shows that regional development was keyed to the contribution of universities first and also to the provision of amenities and services (Mellander and Florida 2006). Ireland's economic success, which is closely related to the flow of foreign direct investment, was primarily the consequence of a ratcheting upward of the supply of technical graduates. Ireland spent little on research and innovation but has the highest number of students in the European Union enrolled in scientific subjects (Crafts 2005).

[13]It is important not to forget the qualifications. Large cities need not be more productive or competitive unless they can muster the industrial, innovative, management, and other capabilities and can create an environment conducive to growth. As a recent study (OECD 2006) indicates, of 44 metro regions, fewer than half exceeded the national average growth rates. On the experience of Latin American cities and the problems they have encountered, see World Bank (2008a).

[14]There is relatively little evidence on city size and air pollution. Overman and Venables (2005) report that in developing countries, both sulfur dioxide and particulates trace an inverted U shape, first rising with size and then falling.

Transport and commuting costs are a separate matter, and for these there are solutions. First, urban intensity can be a plus in that it increases the cost-effectiveness of bus- and rail-based public transport systems, which are also energy efficient—a nontrivial consideration in a world faced with a tightening of energy supplies and a dire need to reduce carbon emissions. Second, an adequate investment in road infrastructure side by side with road-user and parking charges can reduce congestion costs. Third, a polycentric urban system[15] with zoning for mixed use can minimize commuting, distribute traffic more evenly, and prevent peak-time gridlock. Mixed use of buildings increases the efficiency with which urban physical assets are used, while also contributing to the security and social life of neighborhoods.

For middle- and high-income economies that are shifting toward skill-intensive activities, the weight of empirical evidence, especially in the East Asian context, favors the polycentric, large urban regions with a mix of cities (World Bank 2009). They appear to have the edge in terms of competitiveness, knowledge generation, new starts, and productivity. Evidence from Asia also points to the advantages enjoyed by large cities. For instance, a doubling of workers in any one sector leads to a 3 percent increase in gross output per worker, and value added per worker increases by 8 percent (Lee and Zang 1998; see also Becker, Williamson, and Mills 1992; Lall, Shalizi, and Deichmann 2004; Mills and Becker 1986). Indeed, the main drivers of East Asian economies are the large cities, typically dominating in terms of both population and economic size relative to other cities (see tables 6.1 and 6.2). Because of China's size with respect to population and the land mass, the shares of Beijing and Shanghai are much smaller when compared with other East Asian economies, but individually each of these cities is enormous and is the beneficiary of agglomeration economies.[16] If one takes the urban region approach, the

Table 6.1 City Population and Share of National Population, 2005

Attribute	Beijing, China	Shanghai, China	Tokyo, Japan	Seoul, Rep. of Korea	Singapore	Bangkok, Thailand
Size (thousands of people)	15,380	13,603	12,568	10,297	4,266	5,659
Share of national population (%)	1.18	1.04	9.84	21.32	100.00	8.98

Sources: Beijing: Beijing Municipal Bureau of Statistics and Beijing General Team of Investigation under the National Bureau of Statistics 2005; Shanghai: Shanghai Municipal Statistical Bureau 2005; Tokyo: Ministry of Internal Affairs and Communications, Statistics Bureau (http://www.stat.go.jp/english/); Seoul: Seoul Metropolitan Government 2005; Singapore: Singapore Department of Statistics (http://www.singstat.gov.sg/); Bangkok: Bangkok Metropolitan Administration 2005.

---

[15]Polycentric regions can benefit from synergies and external economies; however, these benefits are a function of the degree of complementarity between the different parts and the closeness of interjurisdictional cooperation. The degree to which the benefits from polycentricity have been realized in the Randstad region of the Netherlands is discussed by Meijers (2005). See also Bertaud and others (2007) on land use and polycentricity in Chinese cities.

[16]So also are large metropolitan regions in the United States.

Table 6.2 Cities' Share of National GDP, 1985–2005

City	1985	1990	1995	2000	2005
Beijing, China	2.1	2.7	2.4	2.8	3.7
Shanghai, China	5.2	4.1	4.2	5.1	5.0
Tokyo, Japan	—	17.6	16.1	16.6	18.3
Seoul, Rep. of Korea	24.9	25.3	24.9	24.0	22.8
Singapore	100.0	100.0	100.0	100.0	100.0
Bangkok, Thailand	35.9	40.5	39.1	36.3	44.0

Sources: Beijing: Beijing Municipal Bureau of Statistics and Beijing General Team of Investigation under the National Bureau of Statistics 2005; Shanghai: Shanghai Municipal Statistical Bureau 2005; Tokyo: Tokyo Metropolitan Government, Bureau of General Affairs (http://www.toukei.metro.tokyo.jp); Seoul: Korea National Statistical Office (http://www.nso.go.kr/eng/ and http://global.seoul.go.kr/global/view/business/bus01_02.jsp); Singapore: Singapore Department of Statistics (http://www.singstat.gov.sg/); Bangkok: Bangkok Metropolitan Administration various years and Wikipedia.
Note: — = not available.

Bohai Basin (areas surrounding Beijing), the Pearl River Delta, and the Yangtze River Delta (centered around Shanghai) generate close to half of China's gross domestic product (GDP) even though those areas account for less than 20 percent of the land mass (World Bank 2009). In the smaller countries, the major cities account for a substantial share of the national population and economic activities.

In the future, this advantage is likely to widen as the greater relative scarcity of energy and water encourages urban scale and densities rather than a more dispersed pattern of urbanization and the connecting transport, water, and sanitation infrastructure such a pattern necessitates. Experience—which highlights the role of small and medium-size cities and of "localization economies" arising from urban specialization in single manufacturing industries scattered around the country as in the Russian Federation and the United States—might be increasingly less relevant. Instead, large core cities that benefit from urbanization economies and spawn specialized satellite edge cities enjoying localization economies might be the more efficient model. The closer the satellite cities are to the core city, the larger the mutual gains will be (Rice, Venables, and Patacchini 2006; World Bank 2009; Zheng 2007).

Third, the growth and size of urban agglomerations can also influence the maturing of firms. According to Degroof and Roberts (2004: 4), "academic spin-off ventures in regions outside established high tech clusters tend to stay small 'boutiques.' They fail to grow to become global leaders in their market, in contrast to some of the spin-off firms that have emerged in established USA high tech clusters such as Boston and Silicon Valley. This is a problem that has been observed among European new technology-based firms in general."

Because Malaysia's future industrial competitiveness will be buttressed by both productivity and innovativeness, the urban factor will loom large. Industrial upgrading will depend on dynamic urban centers of a certain scale with the characteristics that enhance agglomeration effects and innovation. Tables 6.3, 6.4, and 6.5 indicate the population, economic size, and foreign direct investment (FDI)

Table 6.3 Population and Its Growth Rates in Four Cities in Malaysia

City	1996	2001	2006
Number of people			
Johor	2,503,903	2,826,548	3,170,516
Kuala Lumpur	1,342,525	1,445,989	1,579,985
Malacca	597,338	660,577	725,323
Penang	1,237,522	1,362,614	1,492,429
Share of national population (%)			
Johor	12.0	12.0	12.3
Kuala Lumpur	6.4	6.2	6.1
Malacca	2.9	2.8	2.8
Penang	5.9	5.8	5.8
Annual population growth (%)			
Johor	n.a.	2.3	2.2
Kuala Lumpur	n.a.	2.1	1.5
Malacca	n.a.	2.2	1.7
Penang	n.a.	2.2	1.6

Sources: Economic Planning Unit of Malaysia; World Bank World Development Indicators database.
Note: n.a. = not applicable.

Table 6.4 GDP and Its Growth Rates in Four Cities in Malaysia

City	1995	2000	2005
GDP (RM million in 1987 prices)			
Johor	18,153.0	23,798.0	29,801.0
Kuala Lumpur	21,157.0	25,963.0	30,412.0
Malacca	5,080.0	6,040.0	7,302.0
Penang	13,293.0	17,064.0	21,277.0
GDP (as a % of national GDP)			
Johor	10.9	11.3	11.6
Kuala Lumpur	12.7	12.4	11.8
Malacca	3.1	2.9	2.8
Penang	8.0	8.1	8.3
Annualized growth rate (%)			
Johor	n.a.	4.6	3.8
Kuala Lumpur	n.a.	3.5	2.7
Malacca	n.a.	2.9	3.2
Penang	n.a.	4.3	3.8

Source: Economic Planning Unit of Malaysia.
Note: n.a. = not applicable.

growth of Malaysia's four largest centers. In terms of the sheer size, the state of Johor leads the pack, with a population of nearly 3.2 million people (12.3 percent of the national share), followed by Kuala Lumpur with 1.6 million and Penang with 1.5 million. FDI is highest in Johor because of its proximity to Singapore, followed by Penang.[17] For the purpose of a knowledge economy, Kuala Lumpur

---

[17]The analysis by Rasiah (2007) indicates that Penang has achieved a higher level of development of the semiconductor and components industries than has Johor because of its integration with the global economy and the presence of many firms in the value chain. However, neither of the urban centers yet has sufficient technological capability to engage in knowledge-intensive activities.

Table 6.5 FDI and Its Growth Rates in Four Cities in Malaysia

City	2000	2003	2006
FDI (RM million)			
Johor	1,861.7	1,108.7	5,488.8
Kuala Lumpur	66.1	38.2	380.3
Malacca	894.6	3,941.4	1,047.6
Penang	3,564.5	1,455.8	3,918.2
FDI (as a % of national FDI)			
Johor	9.4	5.6	27.1
Kuala Lumpur	0.3	0.2	1.9
Malacca	4.5	19.8	5.2
Penang	18.0	7.3	19.4
Annual FDI growth (%)			
Johor	n.a.	−26.7	−4.0
Kuala Lumpur	n.a.	2.5	2,191.2
Malacca	n.a.	660.8	61.1
Penang	n.a.	−26.7	0.3

Source: Economic Planning Unit of Malaysia.
Note: n.a. = not applicable.

has the edge in terms of size, the diversity of services it offers, the presence of research institutions, the headquarters of firms, and the quality of the information technology (IT) infrastructure.[18]

The data on entry of firms among these four cities show that Kuala Lumpur is the leader, followed by Johor, and then very closely by Penang, with Malacca trailing the others in fourth place. Three points emerge from tables 6.6 to 6.9. First, in the Kuala Lumpur–Klang Valley region, the number of net entrants into manufacturing over the most recent three-year period was 1,123, almost four times that in Johor. Second, the number of net entrants has remained relatively stable between 1995 and 2006 in Kuala Lumpur but has declined in the three other urban areas, most steeply in Johor. Third, the industries attracting the new entrants in Kuala Lumpur–Klang Valley are in low- and medium-tech categories—mainly printing, food processing, apparel, plastic, and metal working. Johor is much the same, with greater entry into wood products and furniture, food, textiles, and apparel, which are the industries attracting new firms in Malacca also. Only Penang differs somewhat, with machinery and electronics firms in second and third place, after food, followed by apparel, footwear, plastic products, and printing.

---

[18] The development of biotechnology, defense, and IT-related industries around the Washington, D.C., area can be attributed to the existence of key national research institutes (such as the National Institutes of Health), the federal government, and other regulatory agencies (Feldman 2001; Feldman and Francis 2003).

Table 6.6 Top 10 Net New Entrants in Manufacturing: Kuala Lumpur, 1995–2006

	1995–98			1999–2002			2003–06	
5-digit IC	IC description	Aggregate net entry	5-digit IC	IC description	Aggregate net entry	5-digit IC	IC description	Aggregate net entry
30000	NC	286	30000	NC	265	30000	NC	202
34210	NC (3-digit-level printing, publishing, and allied industries)	120	34210	NC (3-digit-level printing, publishing, and allied industries)	145	34210	NC (3-digit-level printing, publishing, and allied industries)	195
35600	Plastic products, not elsewhere classified	110	31100	NC (3-digit-level food manufacturing)	87	31100	NC (3-digit-level food manufacturing)	169
37100	NC (4-digit-level iron and steel, basic industries)	85	35600	Plastic products, not elsewhere classified	82	34220	NC (3-digit-level printing, publishing, and allied industries)	131
38299	Machinery and equipment, not elsewhere classified	76	34220	NC (3-digit-level printing, publishing, and allied industries)	79	32210	NC (3-digit-level wearing apparel, except footwear)	113
34220	NC (3-digit-level printing, publishing, and allied industries)	74	37100	NC (4-digit-level iron and steel, basic industries)	63	35600	Plastic products, not elsewhere classified	79
32210	NC (3-digit-level wearing apparel, except footwear)	72	32210	NC (3-digit-level wearing apparel, except footwear)	58	37000	NC (2-digit-level basic metal industries)	70
38000	NC (2-digit-level fabricated metal products, machinery, and equipment)	65	37000	NC (2-digit-level basic metal industries)	45	37100	NC (4-digit-level iron and steel, basic industries)	62
37000	NC (2-digit-level basic metal industries)	63	38299	Machinery and equipment, not elsewhere classified	44	38439	Motor vehicles parts and accessories	53
34200	Printing, publishing, and allied industries	60	34200	Printing, publishing, and allied industries	40	31000	NC (2-digit-level food, beverages, and tobacco)	49
		1,011			908			1,123

Source: Economic Planning Unit of Malaysia.
Note: IC = industrial classification; NC = not classified.

Table 6.7 Top 10 Net New Entrants in Manufacturing: Johor, 1995–2006

1995–98			1999–2002			2003–06		
5-digit IC	IC description	Aggregate net entry	5-digit IC	IC description	Aggregate net entry	5-digit IC	IC description	Aggregate net entry
35600	Plastic products, not elsewhere classified	96	33000	NC (2-digit-level wood and wood products, including furniture)	72	30000	NC	74
30000	NC	72	35600	Plastic products, not elsewhere classified	68	31100	NC (3-digit-level food manufacturing)	73
33200	Furniture and fixtures, except primarily metal	71	38000	NC (2-digit-level fabricated metal products, machinery, and equipment)	52	35600	Plastic products, not elsewhere classified	38
32210	NC (3-digit-level wearing apparel, except footwear)	57	38329	Semiconductors and other electronic components and communication equipment and apparatus	32	33000	NC (2-digit-level wood and wood products, including furniture)	22
33000	NC (2-digit-level wood and wood products, including furniture)	41	31100	NC (3-digit-level food manufacturing)	32	34210	NC (3-digit-level printing, publishing, and allied industries)	21
38000	NC (2-digit-level fabricated metal products, machinery, and equipment)	40	31000	NC (2-digit-level food, beverages, and tobacco)	30	33200	Furniture and fixtures, except primarily metal	19
38329	Semiconductors and other electronic components and communication equipment and apparatus	35	30000	NC	29	39090	NC (4-digit-level manufacturing industries, not elsewhere classified)	19
39000	NC (3-digit-level other manufacturing industries)	35	37000	NC (2-digit-level basic metal industries)	25	38410	Shipbuilding and repairing	15
37000	NC (2-digit-level basic metal industries)	35	34210	NC (3-digit-level printing, publishing, and allied industries)	25	39000	NC (3-digit-level other manufacturing industries)	13
31000	NC (2-digit-level food, beverages, and tobacco)	31	39000	NC (3-digit-level other manufacturing industries)	17	32210	NC (3-digit-level wearing apparel, except footwear)	13
		513			382			307

Source: Economic Planning Unit of Malaysia.
Note: IC = industrial classification; NC = not classified.

Table 6.8 Top 10 Net New Entrants in Manufacturing: Penang, 1995–2006

1995–98			1999–2002			2003–06		
5-digit IC	IC description	Aggregate net entry	5-digit IC	IC description	Aggregate net entry	5-digit IC	IC description	Aggregate net entry
30000	NC	81	30000	NC	68	31100	NC (3-digit-level food manufacturing)	48
39000	NC (3-digit-level other manufacturing industries)	62	38000	NC (2-digit-level fabricated metal products, machinery, and equipment)	67	38000	NC (2-digit-level fabricated metal products, machinery, and equipment)	39
35600	Plastic products, not elsewhere classified	52	39000	NC (3-digit-level other manufacturing industries)	61	38329	Semiconductors and other electronic components and communication equipment and apparatus	38
38329	Semiconductors and other electronic components and communication equipment and apparatus	36	35600	Plastic products, not elsewhere classified	37	32210	NC (3-digit-level wearing apparel, except footwear)	31
38299	Machinery and equipment, not elsewhere classified	29	38329	Semiconductors and other electronic components and communication equipment and apparatus	25	35600	Plastic products, not elsewhere classified	26
38000	NC (2-digit-level fabricated metal products, machinery, and equipment)	27	34200	Printing, publishing, and allied industries	21	30000	NC	26
32210	NC (3-digit-level wearing apparel, except footwear)	25	37000	NC (2-digit-level basic metal industries)	20	39000	NC (3-digit-level other manufacturing industries)	25
37100	NC (4-digit-level iron and steel, basic industries)	25	38299	Machinery and equipment, not elsewhere classified	19	34210	NC (3-digit-level printing, publishing, and allied industries)	15
37000	NC (2-digit-level basic metal industries)	19	34100	NC (3-digit-level paper and paper products)	18	34200	Printing, publishing, and allied industries	13
38330	Electrical appliances and housewares	17	31100	NC (3-digit-level food manufacturing)	18	37000	NC (2-digit-level basic metal industries)	13
		373			354			274

Source: Economic Planning Unit of Malaysia.
Note: IC = industrial classification; NC = not classified.

Table 6.9 Top 10 Net New Entrants in Manufacturing: Malacca, 1995–2006

1995–98			1999–2002			2003–06		
5-digit IC	IC description	Aggregate net entry	5-digit IC	IC description	Aggregate net entry	5-digit IC	IC description	Aggregate net entry
30000	NC	15	35600	Plastic products, not elsewhere classified	11	30000	NC	14
34120	Containers and boxes of paper and paperboard	12	31100	NC (3-digit-level food manufacturing)	9	31100	NC (3-digit-level food manufacturing)	13
33200	Furniture and fixtures, except primarily metal	11	34120	Containers and boxes of paper and paperboard	6	32210	NC (3-digit-level wearing apparel, except footwear)	6
38329	Semiconductors and other electronic components and communication equipment and apparatus	9	38329	Semiconductors and other electronic components and communication equipment and apparatus	6	32100	NC (3-digit-level textiles)	6
34210	NC (3-digit-level printing, publishing, and allied industries)	9	30000	NC	6	32000	NC (2-digit-level textile, wearing apparel, and leather industries)	4
31211	Ice factories	9	33200	Furniture and fixtures, except primarily metal	6	34220	NC (3-digit-level printing, publishing, and allied industries)	3
31100	NC (3-digit-level food manufacturing)	7	38391	Cables and wires	5	35600	Plastic products, not elsewhere classified	2
35600	Plastic products, not elsewhere classified	6	36991	Cement and concrete products	4	36900	NC (3-digit-level nonmetallic mineral products)	2
38000	NC (2-digit-level fabricated metal products, machinery, and equipment)	5	31000	NC (2-digit-level food, beverages, and tobacco)	3	38330	Electrical appliances and housewares	2
39000	NC (3-digit-level other manufacturing industries)	4	33000	NC (2-digit-level wood and wood products, including furniture)	3	34210	NC (3-digit-level printing, publishing, and allied industries)	2
		87			59			54

Source: Economic Planning Unit of Malaysia.
Note: IC = industrial classification; NC = not classified.

Table 6.10 Distribution of Labor Force by Educational Attainment, 2002 and 2006

	Percentage of labor force	
Level of education	2002	2006
No formal education	5.3	3.8
Primary	23.5	20.7
Secondary	54.5	56.2
Tertiary	16.7	19.2

Sources: Department of Statistics 2003, 2007.

Table 6.11 Share of Firms That Reported Vacancies for Various Occupations, 2007

Occupation	Share (%)
Professional	27.7
Skilled production worker	44.6
Unskilled production worker	51.7

Source: Authors' calculations based on investment climate data.

Quality of Labor

The quality of the labor force is a critical input toward moving to a knowledge economy. The quality of the labor force measured by educational attainment in Malaysia has steadily improved between 2002 and 2006, reflecting the increase in enrollments in all levels of education (see table 6.10). Despite the gradual increase in educational attainment, a majority of firms in Malaysia have identified shortages of particular skills as a major constraint. This shortage is reflected in the wage premium paid to tertiary-level graduates.[19] Across the skill range, many firms until recently reported unfilled vacancies (see table 6.11).[20] Filling professional positions could take on average 5.4 weeks (table 6.12). In Penang, the shortage was more acute, and it could take more than seven weeks to fill certain professional positions.

Table 6.13 lists the types of skills that firms feel are critical to keep up with rapid technological change but that the current crop of workers do not possess. In 2002, English-language proficiency was by far the most critical skill shortage

---

[19] The return on tertiary education is 17.7 percent, much higher than that on primary (4.5 percent) or secondary education (9.5 percent) (World Bank 2005b), although recent preliminary estimates suggest that the wage premium associated with tertiary education is narrowing.

[20] The proportion of firms reporting unfilled vacancies for professionals changed little from the previous round of the investment climate survey for Malaysia.

Table 6.12 Time It Takes to Fill the Position, 2007

Number of Occupation	weeks
Professional	5.4
Skilled production worker	5.3
Unskilled production worker	3.9

Source: Authors' calculations based on investment climate data.

Table 6.13 Most Critical Skill in Shortage, 2002 and 2007

	Percentage of respondents	
Skill	2002	2007
English-language proficiency	47.5	16.5
Professional communication skills	14.0	6.4
Social skills	8.0	2.5
Team-working skills	6.6	5.9
Leadership skills	4.4	5.5
Time management skills	3.8	5.0
Adaptability skills	1.9	4.2
Creativity and innovation skills	4.1	10.0
Numerical skills	1.1	1.6
Problem-solving skills	1.4	4.0
IT skills	4.1	20.4
Technical or professional skills	3.2	18.0

Sources: For 2002, World Bank 2005b. For 2007, authors' calculations based on investment climate data.

identified.[21] By 2007, the concern had shifted toward IT skills and technical or professional skills, followed by English-language proficiency and creativity. It is difficult to relate these needs to developments in the economy during the five intervening years.

The problem is not a shortage of graduates—in fact Malaysia suffers from a graduate unemployment problem, with 12 to 13 percent of tertiary-level degree holders unemployed.[22] Instead, a lack of desired skills (both basic and technical

---

[21] In many countries, the prospects of finding a job can hinge on English-language proficiency (Lim, Rich, and Harris 2008)—particularly when the economy is well integrated into the global market with a substantial presence of MNCs.

[22] In 2005, 60,000 tertiary graduates were unemployed (Lim, Rich, and Harris 2008). In 2002, among the the unemployed, 18.3 percent finished tertiary education. By 2006, the share of tertiary graduates among unemployed had increased to 24.5 percent (Department of Statistics 2003, 2007). About 40 percent of the unemployed graduates majored in the arts, religious studies, and other subjects not directly related to manufacturing activities (World Bank 2005b).

Table 6.14 Most Important Causes of Vacancy

Cause	Percentage of respondents						
	Most important						Least important
Demand for high wages	31.7	13.4	18.3	14.9	13.7	7.5	0.6
Not enough graduates	2.5	6.4	7.3	15.1	22.4	43.0	3.4
Lack of basic skills	19.1	30.3	21.9	16.8	9.9	1.9	0.1
Lack of technical skills	17.2	24.2	25.8	17.9	11.3	3.6	0.0
No applicant (unskilled)	9.5	10.7	16.2	18.2	22.6	21.4	1.5
High turnover	18.8	14.4	11.1	15.7	19.3	19.3	1.3

Source: Authors' calculations based on investment climate data.

skills) is identified as the most important constraint (see table 6.14). This finding suggests that quality of education and possibly skill mismatch are the areas that warrant closer attention. Almost every country bemoans the persistence of skill mismatches and struggles to minimize them. Whether they can be eased in Malaysia will depend on what teaching institutions can do, but also on the wages employers are willing to pay to obtain the quality and skills they claim to be seeking. The reluctance of firms to offer higher wages suggests that although firms complain about skill shortages, their strategies are still based on a low-skill, low-cost model. This tendency may be changing as the pressure to upgrade mounts. From the policy makers' point of view, two broad approaches can better match the skills of workers to the jobs being advertised. One focuses on providing training once students finish formal schooling, and the other on improving the quality of formal education.

Skills Development

Recognizing the need for improving workers' skills and decreasing the high turnover rate, which supports the case for public provision, Malaysia introduced the Human Resource Development Fund (HRDF) in 1992 (Tan and Gill 2000). The HRDF is funded through a payroll levy (1 percent),[23] and firms that have contributed to HRDF for at least six months are eligible to claim a portion of the levy for employee training. The claimable portion varies by skills and by the size of firms. Firms can claim a larger portion of the levy for training for higher skills (technical, supervisory, and so forth), and smaller firms are typically allowed to claim more for the same kind of training relative to large firms (Tan and Gill 2000). An analysis of the HRDF finds that large firms use it more than small firms. When small firms take advantage of this opportunity, their focus is on technical training, especially in subsectors such as textiles, apparel, and footwear. Large firms, in contrast, tend to use HRDF more for retraining existing production workers to keep up with changing production processes. The introduction of HRDF has increased

---

[23] Many countries also adopt similar levies ranging from 1 to 2 percent of the payroll (Ziderman 2003).

training by firms. However, four issues remain. First, a sizable portion of firms—especially smaller firms—still do not contribute to the fund. Second, even if firms are registered and eligible to claim HRDF, one-third of firms do not do so. These nonclaiming firms also provide less training than those that claim the benefit regularly (Tan and Gill 2000; World Bank 2005b). Third, as in many other countries, the responsibilities for vocational and skill training in Malaysia are divided among several different ministries. More coordination among these different entities is essential if skill training is to become more effective (Nexus Associates 2005; Tan and Gill 2000). Fourth, improving the quality of training institutes requires that they be more autonomous and be exposed to greater competition. Moreover, close links with industry (as seen in the Penang Skills Development Centre) are desirable to improve the curriculum and the relevance of these institutes.[24]

However, the empirical literature suggests that training is most effective for those who have a strong primary and secondary education prior to receiving the training (Heckman and LaFontaine 2008). This finding calls for strengthening formal secondary education in parallel with the active labor market policies.

Secondary Education

Relative to other economies in East Asia, the gross enrollment ratios in Southeast Asia are somewhat lower, ranging from 64 percent in Indonesia to 83 percent in the Philippines (see figure 6.1). Among the Southeast Asian economies, the quality of secondary education in Malaysia, as measured by the international test scores, is still the highest. And according to international standardized tests, the capabilities of Malaysian secondary-level students are close to the international average. However, the latest Trends in International Mathematics and Science Study (TIMSS) results indicate that Malaysia is slipping below the international average and still trailing behind the frontrunners (Hong Kong, China; the Republic of Korea; Singapore; and Taiwan, China) (see tables 6.15 and 6.16).

International experience with reforming secondary and primary education offers several pointers, such as improving the qualifications of teachers, recruiting high-caliber students to teach, making the teaching profession more attractive, and introducing more competition among schools, which also requires that greater autonomy be given to schools (Brock, Marshall, and Tucker 2009). Firms often voice dissatisfaction with the technical and soft (noncognitive) skills—communication and teamwork among others—of new hires. The foundations of these skills need to be developed by primary and secondary schools. Raising the standard of English-language skills and training in the sciences would be highly beneficial for students. A study by Lim, Rich, and Harris (2008) found that English proficiency before entering a university has a large bearing on labor

---

[24] The approach adopted by the providers of technical and vocational education offers a telling example.

Figure 6.1 Secondary School Gross Enrollment Ratio, 2006

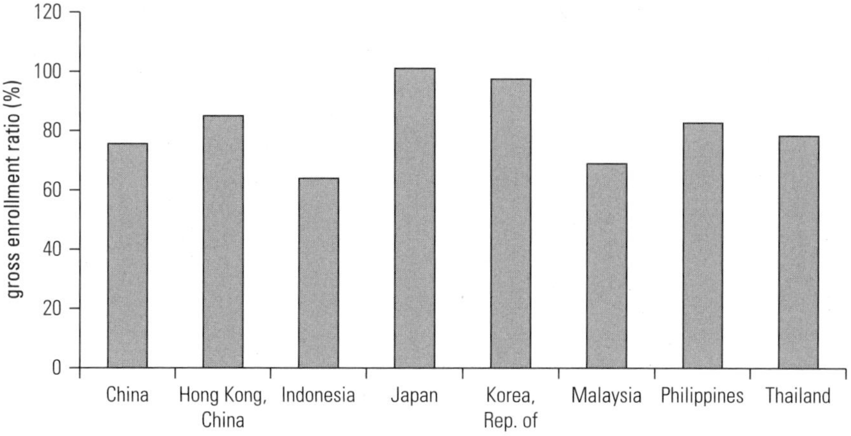

[Bar chart showing gross enrollment ratio (%) for: China ~75, Hong Kong, China ~85, Indonesia ~65, Japan ~100, Korea, Rep. of ~97, Malaysia ~68, Philippines ~83, Thailand ~78]

Source: World Bank World Development Indicators database.
Note: Data for the Republic of Korea are from 2007; data for Malaysia are from 2005.

Table 6.15 Eighth-Grade TIMSS Scores for Mathematics for Selected East Asian Economies, 1999, 2003, and 2007

Economy	1999	2003	2007
Taiwan, China	585	585	598
Korea, Rep. of	587	589	597
Singapore	604	605	593
Hong Kong, China	582	586	572
Japan	579	570	570
United States	502	504	508
International average	487	466	500
Malaysia	519	508	474
Thailand	467	n.a.	441
Indonesia	403	411	397
Philippines	345	378	n.a.

Sources: Gonzales and others 2004, 2008; Mullis and others 2000.
Note: n.a. = not applicable. Economies are ranked by their score in 2007.

market outcomes after graduation, but efforts to improve English-language skills during university years are inconsequential. This finding suggests that students should be exposed to instruction in English at the secondary level, if not earlier. Pedagogy can also be modified to include more team-working activities, and in addition to drilling students in techniques and stressing fact memorization, schools should encourage efforts to nurture creativity.

Table 6.16 Eighth-Grade TIMSS Scores for Science for Selected East Asian Economies, 1999, 2003, and 2007

Economy	1999	2003	2007
Singapore	568	578	567
Taiwan, China	569	571	561
Japan	550	552	554
Korea, Rep. of	549	558	553
Hong Kong, China	530	556	530
United States	515	527	520
International average	488	473	500
Malaysia	492	510	471
Thailand	482	n.a.	471
Indonesia	435	420	427
Philippines	345	377	n.a.

Sources: Gonzales and others 2004, 2008; Mullis and others 2000.
Note: n.a. = not applicable. Economies are ranked by their score in 2007.

Quality of Tertiary Education

Malaysian universities have begun emphasizing the quality of teaching and research, and this effort is reflected in modest gains in their international rankings in 2007 and 2008 (see table 6.17).[25] Top universities in the region are mainly concentrated in Northeast Asia. Overall, Malaysian universities now rank among the top 200 to 300 universities worldwide and among the top 30 to 50 in East Asia. Clearly, plenty of scope for further progress remains.

The evaluation of the tertiary education system in Malaysia has identified four principal shortcomings. These shortcomings are in the areas of governance (autonomy, financing, and accountability); quality (especially that of faculty and also the quality assurance system); relevance of skills and knowledge instilled in students; and links with firms and integration in the nascent national innovation system (World Bank 2007c). Much like other Southeast Asian countries, Malaysia is facing both quality issues and a rising demand for tertiary education. Attaining both goals will require additional public funding for research. In addition, the government will need to encourage the establishment of private institutions to supplement the public university sector. Increasing cost sharing would facilitate the expansion of the tertiary education sector and improve the quality of schooling.[26]

---

[25]There was no Malaysian university listed in the ranking published by Shanghai Jiao Tong University.

[26]Such cost sharing will invariably include student fees. The government needs to assess whether tuition fees and other expenditures associated with tertiary education have an adverse effect on equity in access and which types of financial aid and grants can be used to address the equity concern (World Bank 2007c).

Table 6.17 Ranking of Selected Universities in East Asia, 2007 and 2008

International rank 2008	International rank 2007	Regional rank 2008	School	Economy
19	17	1	University of Tokyo	Japan
25	25	2	Kyoto University	Japan
26	18	3	University of Hong Kong	Hong Kong, China
30	33	4	National University of Singapore	Singapore
39	53	5	Hong Kong University of Science and Technology	Hong Kong, China
42	38	6	Chinese University of Hong Kong	Hong Kong, China
44	46	7	Osaka University	Japan
50	36	8	Peking University	China
50	51	9	Seoul National University	Korea
56	40	10	Tsinghua University	China
166	223	23	Chulalongkorn University	Thailand
230	**246**	**33**	**Universiti Malaya**	**Malaysia**
250	**309**	**35**	**Universiti Kebangsaan Malaysia**	**Malaysia**
251	284	36	Mahidol University	Thailand
276	398	39	University of the Philippines	Philippines
287	395	41	University of Indonesia	Indonesia
313	**307**	**43**	**Universiti Sains Malaysia**	**Malaysia**
315	369	44	Bandung Institute of Technology	Indonesia
316	360	45	Universitas Gadjah Mada	Indonesia
320	**364**	**46**	**Universiti Putra Malaysia**	**Malaysia**
356	**401–500**	**51**	**Universiti Teknologi Malaysia**	**Malaysia**

Source: Times Higher Education Supplement
(http://www.topuniversities.com/university_rankings/results/2008/overall_rankings/fullrankings/).

The key is to recognize that not all universities can or should be research oriented. The tertiary education system needs to be differentiated. In this regard, the Malaysian government has taken the right steps and has identified four universities that would do the lion's share of research. However, raising the quality of research goes hand in hand with competition for R&D funding and requires that efforts be made to promote the establishment of centers of excellence that respond to the national, regional, and societal needs.

The experience of aligning the demand and supply of skills in other countries can offer a few concrete suggestions. Actively soliciting firms' inputs for curriculum development can yield positive results. Other possible approaches toward enhancing the relevance of skills and knowledge taught at universities include internships to gain up-to-date practical knowledge, faculty involvement in technical assistance to firms, and continuing tracer studies to track the careers of graduates.

Furthermore, the quality of tertiary education in Malaysia would benefit from a number of initiatives such as

- Simplifying the review of academic standards
- Ensuring the independence of quality assurance bodies by applying quality assurance schemes impartially to public and private universities
- Offering more courses in English, which is in high demand by students and by firms

These measures will need to be complemented by steps to raise the quality of faculty by

- Seeking a better balance among the teaching, research, and administrative functions performed by the faculty
- Using student evaluations of teachers in making tenure and promotion decisions
- Involving senior researchers and faculty members in instructing undergraduate classes so as to link advanced research to teaching
- Relaxing the retirement age of faculty members (currently 56) so as to counter the faculty shortage, assuming of course that those on the verge of retirement remain qualified to teach

Tertiary Institutes and Research Activities

The number and scale of tertiary institutions in an urban area are important determinants of the supply of technical skills, and the degree to which these bodies offer graduate education and engage in research contributes to the knowledge capital of a city. In this regard, Kuala Lumpur leads other cities in Malaysia.

The majority of Malaysia's government research institutes are located in Kuala Lumpur—20 in all, including the leading ones: MIMOS (Malaysian Institute of Microelectronic Systems), SIRIM (Standards and Industrial Research Institute of Malaysia), and the Palm Oil Board (see table 6.18). Including the 10 research centers located within Putrajaya, Kuala Lumpur has 30 research institutes. In contrast, there are only two in Penang (the Fisheries Research Institute and the fledgling Malaysian Institute of Pharmaceuticals and Nutraceuticals at Universiti Sains Malaysia, which is tasked with conducting research on traditional herbal medicines); two in Johor (the Malaysia Pineapple Industry Board and Pusat Penyelidikan Ternakan Air Payau); and one in Malacca (Jabatan Perikanan).

Table 6.18 Number of Government Research Institutes, 2008

	City				
	Johor	Kuala Lumpur	Malacca	Penang	Putrajaya
Number of government research institutes	2	20	1	2	10

Source: Economic Planning Unit of Malaysia.

Table 6.19 Location of Public and Private Universities, 2008

Type of university	City				
	Cyberjaya	Johor	Kuala Lumpur	Malacca	Penang
Public	0	2	3	1	1
Private	2	0	8	0	0

Source: Economic Planning Unit of Malaysia.
Note: Only the main campuses are included in this table.

Most of the universities, especially the private universities, are located in Kuala Lumpur, a factor that strengthens that region's knowledge capacity (see table 6.19). Other cities rely on a public university or two for tertiary education, such as Universiti Sains Malaysia (USM) in Penang, Universiti Teknologi Malaysia (UTM) and Universiti Tun Hussein Onn Malaysia in Johor, and Universiti Teknikal Malaysia Melaka in Malacca.

The leading universities are making an effort to expand their postgraduate and postdoctoral programs and to promote faculty research. The scale of the R&D conducted is greatest in Kuala Lumpur, but universities in other cities are also entering the fray. UTM was granted eight patents between 1995 and 2005 and ranks at the top, as measured by R&D spending by universities. In addition, UTM leads in R&D spending, mainly in engineering fields. USM was granted 10 patents between 1995 and 2005 (see table 7.14 in chapter 7). It ranks second among universities in R&D spending (see table 7.11 in chapter 7). USM's R&D emphasizes biochemistry, energy technology, and process technology and engineering (see table 7.10 in chapter 7). Universiti Teknologi MARA ranked seventh among universities in terms of R&D spending. The bulk of its research effort is oriented toward manufacturing and production engineering. However, the total R&D spending of RM 20 million is only one-fifth of the spending done by the other major research universities.

Notwithstanding the scale of ongoing activities, recent efforts to spur research are thin and scattered in Malaysia, as in the other Southeast Asian countries, in large part because postgraduate and postdoctoral programs in Southeast Asia, where they exist at all, are exceedingly small and diffuse. Few incentives exist for local or foreign researchers to engage in research in Malaysian universities, and few do. There are ambitious plans to dramatically increase postgraduate studies; whether they succeed will be the function of a sustained injection of resources and a rising, well-compensated demand for higher-level skills and academic research.

Role of Large Firms

East Asian experience indicates that the acceleration of technological upgrading and the strengthening of innovation capabilities have been associated with the emergence of a few large and dynamic multinational firms. Whether in Japan; India; Korea; or Taiwan, China, the strategies pursued by these firms have been broadly similar. They have involved early technological borrowing from abroad and consolidation of production capabilities (for manufactures or services), followed by determined efforts to acquire design and innovation capabilities (through licensing and takeover of foreign firms). Such efforts have been backed by heavy investment in R&D, which combined the advantage of cost-efficiency with innovativeness.[27] Now Chinese firms such as China International Marine Containers, Huawei, Chery, BYD Auto, and Wanxiang are proceeding down this very same path (Zeng and Williamson 2006). No Malaysian manufacturing firm has established itself as a major contract (original equipment manufacturer) supplier of a product or a service with an expanding international market. Firms such as Eng Teknologi and Pentamaster have acquired original design manufacturer status, but there are few of these firms, and they remain relatively small regional suppliers. This situation might be related not just to firm-level strategies, choice of products, and determination to emerge as world-class players, but also to the size of the local economy and how it affects productivity and relevant capabilities. Firms in both Kuala Lumpur and Penang remain small in scale, as are even the largest Malaysian nonfinancial firms, with the exception of Petronas.

Large firms play a significant role in the development of local industries, because they can become a fertile source of new spinoff firms. Their applied research, worldwide brand names, and marketing strengths can support a strategy in which innovation is a major strand, and they are frequently critical to the future of new products and services introduced by start-ups. In Silicon Valley, Fairchild Semiconductor, a prototypical axial firm, served as the launching pad for scores of start-ups.[28] In San Diego, Hybritech (see figure 6.2) and Linkabit were responsible for a number of start-up firms in the biotechnology–pharmaceutical and wireless telecommunication clusters. Similarly in Cambridge, United Kingdom, Cambridge

---

[27]The successful and strongly government-supported efforts of firms from Taiwan, China, are described by Mathews and Cho (2000). The Korean experience is detailed by Linsu Kim (1997).

[28]Fairchild Semiconductor was formed by eight engineers from Shockley Semiconductor in 1957. These eight engineers, called the "Traitorous Eight" by William Shockley, were Julius Blank, Victor Grinich, Jean Hoerni, Eugene Kleiner, Jay Last, Gordon Moore, Robert Noyce, and Sheldon Roberts. Fairchild Semiconductor itself was responsible for spawning 65 other firms, often referred to as "Fairchildren." Among them are Intel, Amelco, Advanced Micro Devices, LSI Logic, Altera, Xilinx, and other semiconductor firms. In addition to spinning off these firms, Fairchild was the progenitor of venture capital firms such as Kleiner Perkins Caulfield & Byers, which helped to finance Amazon, Google, and Sun Microsystems and catalyzed the development of semiconductor clusters in northern California. Another important venture capital firm in Silicon Valley, Sequoia Capital, is also one of the Fairchildren. When Fairchild was formed, no venture capital was available on the West Coast, and the eight engineers needed to approach Fairchild Camera and Instrument for a loan (with the option to purchase the firm). When Fairchild was established, the semiconductor industry was at a very early stage of development, and

(*continued on next page*)

Consultants (CCL) is responsible, directly and indirectly, for a host of daughter firms. Some of these start-ups spun off from CCL, whereas others have close social ties to CCL.[29] The role of anchor firms is important even in lesser known clusters. For instance, the majority of firms in the wireless telecommunications cluster in Denmark can be traced back to a single firm, S.P. Radio, which grew initially from its sales and manufacturing of consumer radios and later of maritime radio equipment (Dahl, Pedersen, and Dalum 2005). The emergence of Akron, Ohio, as the major cluster for automotive tire production is because of the establishment of BFGoodrich in 1896 (Buenstorf and Klepper 2009).[30]

Are there Malaysian firms with the managerial expertise, the scale, and the human and research resources to play such a role? Table 6.20 lists some of the top 500 firms listed on the Kuala Lumpur Stock Exchange. Many of these firms are banks, which can assist with the birth and growth of firms, especially when the venture capital industry is still in its infancy, although banks are frequently ill equipped to coach and mentor start-ups and help them find the right contacts in the way that experienced venture capitalists can. On the list are a number of firms with agricultural operations, reflecting the legacy of the plantation economy. Only a handful of firms are in the manufacturing industry. Notably, there are also a number of firms in engineering and R&D services.

More than 60 percent of these firms are located in Kuala Lumpur and Putrajaya (see table 6.21).[31] Ten firms are headquartered in Selangor. Only two firms are headquartered in Penang (Oriental Holdings in the automotive sector and Uchi Technologies in electronic components), and none are in Johor and Malacca. In other words, the large and medium-size firms that could accelerate technological catch-up and serve as the kernel of new clusters are virtually all located in

---

(*continued from previous page*)

there were no equipment manufacturers or parts makers. Fairchild Semiconductor needed to manufacture almost everything in house. This situation paved the way for the development of other firms in the area to supply materials and equipment to Fairchild (Nuttall 2007). Netscape Communications also was an influential firm, which spawned Barksdale Group, Epinions, Backflip, Friendster, Tellme, Kontiki, Hopelink, OurPictures, Viralon, Meer.net, RxCentric, MedicaLogic, Cloudmark, Xoopit, Responsys, Mozilla Foundation, and Sherpalo Ventures (Takahashi 2007).

[29] Firms related to CCL include Domino Printing Sciences, P.A. Technologies, and Scientific Generics, among others (Vyakarnam 2007).

[30] BFGoodrich started as a tire manufacturer for bicycles and began manufacturing automobile tires when the automobile industry took off. The proximity of Akron to Detroit, Michigan, was also an advantage (Buenstorf and Klepper 2009).

[31] In the United States, 33 percent of all headquarters are located in the top five largest metropolitan areas in 2000. Although the trend is such that the headquarters of manufacturing firms are dispersing away from the largest metropolitan areas, headquarters of services firms (especially high value-adding firms) are concentrating in the largest cities (Diacon and Klier 2003).

Figure 6.2 Hybritech and Its Daughter Firms in San Diego

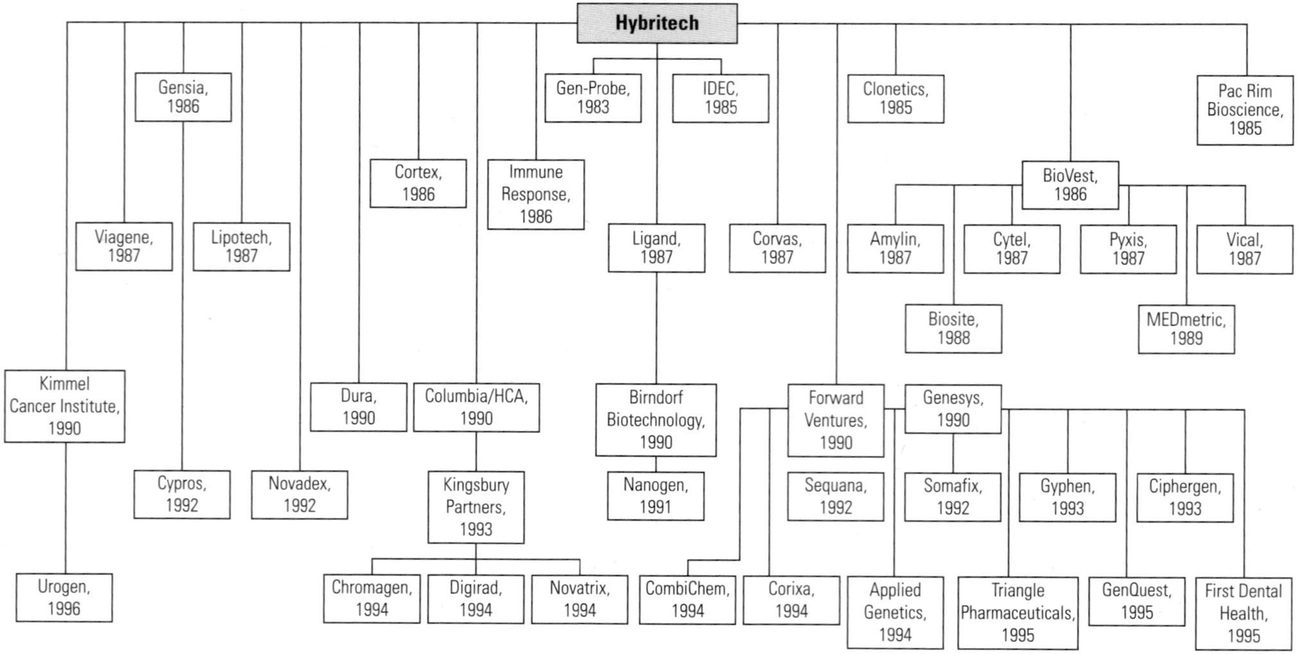

Source: Smilor and others 2005.

Table 6.20 Malaysia's 40 Largest Firms with Spinoff Potential, 2007

Ticker	Rank	Firm	Industry	Market capitalization (US$ million)
MAY	1	Malayan Banking	Commercial banks, non-U.S.	12,576.7
BCHB	3	Bumiputra-Commerce Holdings	Commercial banks, non-U.S.	9,735.6
PBKF	7	Public Bank–Foreign	Commercial banks, non-U.S.	9,411.2
PBK	8	Public Bank	Commercial banks, non-U.S.	9,159.5
IOI	9	IOI Corporation	Agricultural operations	8,367.4
SDY	11	Sime Darby	Diversified operations	6,531.7
SIDBY	14	Sime Darby–ADR	Diversified operations	5,172.9
GIL	15	Genting International	Real estate operations and development	4,443.0
MMC	24	MMC Corporation	Diversified operations	3,048.3
HLBK	25	Hong Leong Bank	Commercial banks, non-U.S.	2,508.3
PEP	28	PPB Group	Food, miscellaneous and diversified	2,221.0
UMWH	29	UMW Holdings	Auto and truck parts and equipment, original	2,067.3
ASTR	32	Astro All Asia Networks	Multimedia	1,914.0
KGB	37	Kumpulan Guthrie	Agricultural operations	1,657.6
NESZ	38	Nestlé Malaysia	Food, miscellaneous and diversified	1,609.7
SPSB	41	S P Setia	Real estate operations and development	1,317.8
IOIP	51	IOI Properties	Real estate operations and development	1,200.6
BAK	55	Batu Kawan	Chemicals, specialty	978.8
ORH	60	Oriental Holdings	Automotive, cars and light trucks	887.2
KLCC	61	KLCC Property Holdings	Real estate management and services	887.0
UPB	63	United Plantation Berhad	Agricultural operations	848.0
PROH	64	Proton Holdings	Automotive, cars and light trucks	793.3
TTNP	67	Titan Chemicals	Chemicals, plastics	731.9
UPL	70	United Plantation	Agricultural operations	714.4
TOPG	82	Top Glove Corp.	Rubber and vinyl	562.8
WCT	101	WCT Engineering	Engineering and R&D services	459.9
SGB	108	Scomi Group	Oil field machinery and equipment	384.0
RANH	118	Ranhill	Engineering and R&D services	362.2
PETR	124	Petra Perdana	Engineering and R&D services	332.8
GRPB	128	Green Packet	Internet Infrastructure software	331.7
MUHI	132	Muhibbah Engineering	Engineering and R&D services	315.0
UCHI	137	Uchi Technologies	Electronic components, miscellaneous	311.1
GOLD	174	Goldis	Medical products	212.7
STEM	188	StemLife	Medical labs and testing services	191.6
KRI	190	Kossan Rubber Industries	Rubber and plastic products	198.4
KPJ	207	KPJ Healthcare	Medical, hospitals	169.4
VADS	217	VADS	Computers, integrated systems	150.8
MDJ	242	Mudajaya Group	Engineering and R&D services	122.4
N2N	243	N2N Connect	R&D	128.3
SAAG	288	SAAG Consolidated	Oil field machinery and equipment	108.3

Source: Economic Planning Unit of Malaysia.
Note: The capitalization cutoff used to constitute this list of "large firms" was US$100 million.

Table 6.21 Locations of Headquarters

	City					
	Ipoh	Kuala Lumpur	Penang	Perak	Putrajaya	Selangor
Number of firms	1	23	2	2	2	10

Source: Economic Planning Unit of Malaysia.
Note: Figures are based on the locations of headquarters of firms in table 6.20.

Kuala Lumpur. Most major Indonesian and Thai firms are also headquartered in the capital cities.

From table 6.21, it is apparent that Kuala Lumpur is the business center of Malaysia and the hinge of the knowledge economy, just as Bangkok is the hinge of Thailand's fledging knowledge economy. Most of the leading firms prefer to locate in Kuala Lumpur because it is the only city in Malaysia that offers some measure of agglomeration economies, and because it is adjacent to the administrative center of the country and also to the leading science park. Kuala Lumpur is served by an excellent new airport (Kuala Lumpur International Airport at Sepang) and is well furnished with other amenities. Moreover, most of Malaysia's leading universities and research institutes are in the Kuala Lumpur–Klang Valley area, which further enhances the attraction of this urban region to firms and increases the chances of Kuala Lumpur entering a virtuous economic spiral.

The listing also points to the absence thus far in Malaysia and other Southeast Asian countries of firms that have been able to assemble the expertise in their field, to conduct research, and to innovate on a scale that would potentially enable them to nucleate industrial clusters as Fairchild and CCL were able to do. The absence of such firms is not a binding constraint, but it does mean that growth of competitive firms with innovative products and services requires alternative sources of stimuli and entrepreneurship.

IT Infrastructure

A meta-analysis by Straub (2007) of economic studies analyzing the link between infrastructure investment and urban development outcomes found that in 63 percent of the studies, infrastructure investment had a positive and significant effect on urban economic growth. For instance, one study found that an increase in telecommunications penetration is responsible for close to one-third of the growth in the Organisation for Economic Co-operation and Development (OECD) over the past 20 years. Korea, for example, has made good use of broadband penetration to promote multimedia-based activities and Web services, especially in Seoul. The IT infrastructure, in particular, is becoming an important factor influencing the efficacy of research efforts. Better IT infrastructure permits researchers to connect globally, to exchange ideas, to search for ideas, to access papers, and so on. In this

Table 6.22 Broadband Subscribers, 2001–07

	Number of subscribers per 1,000 population		
Economy	2001	2004	2007
China	0.3	19.1	50.0
Indonesia	0.1	0.4	1.0
Japan	30.1	153.0	224.7
Korea, Rep. of	164.9	247.9	306.2
Malaysia	0.2	9.9	51.5
Philippines	0.1	1.1	11.0
Singapore	36.6	130.3	198.7
Taiwan, China	50.6	165.3	209.3
Thailand	—	1.2	9.4

Source: International Telecommunication Union.
Note: — = not available.

regard, Malaysia is moving ahead, but there is still room for catching up. Although the number of people subscribing to broadband has increased rapidly in Asia since 2001, diffusion of broadband in Malaysia is still low (see table 6.22) relative to OECD countries, although 52 people per 1,000 were connected in 2007 (see table 6.23). Broadband is offered through four Internet Service Providers in Malaysia. The connection speeds they offer range from 128 kilobits per second to 2 megabits per second (see table 6.24). These speeds are much lower than the average speed available in a number of countries listed in table 6.23. Both Japan and Korea have invested heavily in laying fiber-optic cables to make high-speed Internet connections available to residents in major cities. Because of this outlay, their average speeds are 61.0 and 45.6 megabits per second, respectively, much faster than in any other OECD country. Faster speed permits the transfer of larger amounts of data in a shorter period of time, making it possible to collaborate more easily with others on technical projects involving the manipulation of masses of data.[32]

For the purposes of exchanging ideas with parties overseas, the international bandwidth can be used as a measure of connectivity. The United States is the most connected country with connections exceeding 970 gigabits per second (in 2005), followed by many of the OECD countries. Relative to these countries, the connections available in Malaysia are few, and the capacity is a mere 3 gigabits per second (see table 6.25). Although the available capacity permits access to information from overseas or in Malaysia, the amount of information that can be accessed at any given time is limited, and the speed of access is slow because of congestion.

---

[32] In addition, faster connection speeds available to regular consumers can lead to the development of new industries, such as Internet Protocol television and online gaming, and wider usage of existing forms of e-commerce (online banking, shopping, and so forth).

Table 6.23 Information Technology and Innovation Foundation Broadband Rankings, 2006

Rank	Economy	Subscribers per household (%)	Average speed (megabits per second)	Price per month for 1 megabit per second of fastest technology (US$)
1	Korea, Rep. of	90.0	45.6	0.45
2	Japan	52.0	61.0	0.27
3	Finland	57.0	21.7	2.77
4	Sweden	49.0	18.2	0.63
5	United States	51.0	4.8	3.33
6	Switzerland	68.0	2.3	21.71
7	Australia	50.0	1.7	2.39
8	United Kingdom	50.0	2.6	11.02
9	Germany	38.0	6.0	5.20
10	Ireland	37.0	2.2	13.82
	Average	46.0	9.0	16.52

Source: Atkinson 2007.

Table 6.24 Average Speed of Internet Connections Offered in Malaysia, 2008

Broadband services providers	Home	Business
TIMENet	448 kilobits per second	256 kilobits per second–2 megabits per second
JARING Net	512 kilobits per second–1 megabit per second	
TMNet Streamyx	384 kilobits per second–1 megabit per second	1–2 megabits per second
Maxis	384 kilobits per second–2 megabits per second	128 kilobits per second–512 kilobits per second

Source: Economic Planning Unit of Malaysia.

The available capacity in Thailand is 6.8 gigabits per second, twice as much but still low. The Philippines is even with Malaysia, whereas Indonesia brings up the rear with just 1.5 gigabits per second.

Advances in IT and its widespread use in the OECD countries have led to the explosion of new businesses and of Internet-based activities such as e-commerce and social networking sites. IT has galvanized industries that rely on outsourcing from the advanced countries. In this latter respect, the Philippines has emerged as a leader, with a flourishing call center industry and businesses dealing in software, animation, and outsourcing of transcription by medical and legal offices in the United States. New types of firms as well as traditional industries rely on the Internet in their day-to-day operations (ordering, invoicing, tracking, and fulfillment)

Table 6.25 International Bandwidth, 1999–2005

Economy	Bandwidth (megabits per second)		
	1999	2002	2005
Australia	730	10,498	—
China	351	9,380	136,106
Finland	670	16,587	22,617
Germany	11,834	260,668	566,056
India	267	1,870	20,000
Indonesia	120	573	1,507
Ireland	239	13,501	24,587
Japan	2,643	30,286	132,608[a]
Korea, Rep. of	1,251	17,207	49,766
Malaysia	188	1,321	3,193
Netherlands	10,874	167,232	334,578
Philippines	125	891	3,215
Singapore	845	5,898	24,704[a]
Sweden	4,388	94,896	157,636
Switzerland	5,257	65,827	71,464
Thailand	118	1,011	6,808
United Kingdom	18,338	319,663	781,554[a]
United States	27,388	381,693	970,594[a]

Source: International Telecommunication Union.
Note: — = not available.
a. Data are for the year 2004.

for communications and coordination within production networks to improve their efficiency. Thus far, information and communication technology (ICT) capital is contributing more to growth in the advanced countries, whereas in East Asia, investment in physical capital has the edge.[33] Looking ahead, one is likely to see the gap narrow as services activities mature and become internationally competitive (see figure 6.3). The Economist Intelligence Unit compiles e-readiness scores to measure the e-business environment and opportunities. Singapore is the highest ranked, followed by Korea; Taiwan, China; and Japan (see table 6.26).[34] Malaysia ranks after Japan with a considerable gap in the score, followed by

---

[33] See Draca, Sadun, and Van Reenen (2007) for an extensive review of the effect of ICT investment on productivity. The research shows that investment in ICT hardware alone would not be sufficient. To take full advantage of ICT capital, firms also need to transform their organizational structure (Bloom, Sadun, and Van Reenen 2007).

[34] However, see Wong (2008) regarding Singapore's mixed record in providing high-speed Internet access to households and developing Internet content providers.

Figure 6.3 Information and Communication Technology Investment and Labor Productivity Growth, 1989–2005

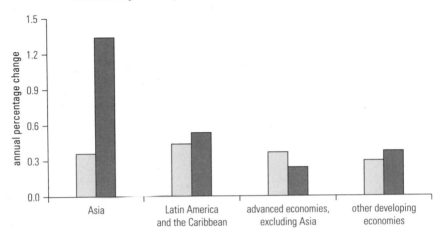

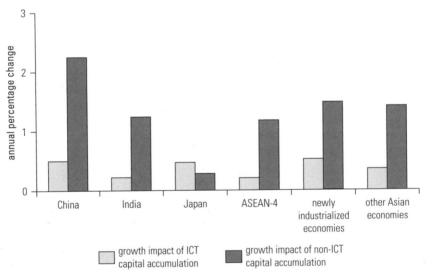

Source: Jaumotte and Spatafora 2007.
Note: ASEAN-4 refers to four members of the Association of Southeast Asian Nations: Indonesia, Malaysia, the Philippines, and Thailand.

Thailand, India, China, and Indonesia. All the economies listed in table 6.26 improved their scores steadily over the years. China and India improved much more than the others, although they started out from lower bases. Again, Malaysia is in the middling range (scores for individual cities in Malaysia were unavailable), Thailand is a little below, and Indonesia occupies a lower rung.

Table 6.26 E-Readiness Scores, 2003–07

Economy	Score				
	2003	2004	2005	2006	2007
Singapore	8.20	8.02	8.18	8.24	8.60
Korea, Rep. of	7.80	7.73	7.66	7.90	8.08
Taiwan, China	7.40	7.32	7.13	7.51	8.05
Japan	7.10	6.86	7.42	7.77	8.01
Malaysia	5.60	5.61	5.43	5.60	5.97
Thailand	4.20	4.69	4.56	4.63	4.91
India	3.90	4.45	4.17	4.25	4.66
China	3.80	3.96	3.85	4.02	4.43
Indonesia	3.30	3.39	3.07	3.39	3.39

Source: Economist Intelligence Unit (http://globaltechforum.eiu.com/index.asp?layout=rich_story&channelid=4&categoryid=29&doc_id=10599).
Note: Economies are ordered by their score in 2007. The rankings evaluate a country's e-business environment and how amenable a market is to Internet-based opportunities. Scores may range from 0 to 10.

The current IT infrastructure in Malaysia and in the other Southeast Asian countries cannot adequately support a more knowledge-intensive growth strategy. Not only do the capacity and speed of Internet connections need to be improved domestically, but also a larger capacity to connect to the rest of the world is required. Such an undertaking will entail more investment, especially in the main urban centers. An increase in international bandwidth is also required for diversifying connections. The earthquake in Taiwan, China, in 2006 highlighted the vulnerability of the international system by disrupting the connectivity to the world for a brief period. Other episodes since then have further underscored this vulnerability. To avoid breakdowns from other causes in the future, Southeast Asia needs to build up international bandwidth and redundancy.

In 1995, the Malaysian government launched the Multimedia Super Corridor (MSC), covering the 15- by 75-kilometer area just outside of Kuala Lumpur, in an attempt to create an IT industry hub in Malaysia (Fleming and Søborg 2002).[35] It offered a modern physical infrastructure (fiber-optic backbones) and a number of fiscal incentives, such as an income tax exemption for up to 10 years, a 100 percent investment tax allowance, free imports of multimedia equipment, and employment of foreign knowledge and IT workers (Fleming and Søborg 2002). With the MSC, the authorities sought to accelerate Malaysia's transition to a knowledge-based economy

---

[35] The area covered by the MSC extending southward from Kuala Lumpur is larger than Singapore (Fleming and Søborg 2002). Within this stretch, the new cities of Putrajaya (the administrative center) and Cyberjaya (the high-tech industrial center) were constructed (Scott 1998). The MSC is anchored by two points: by Kuala Lumpur International Airport in the south and by the Kuala Lumpur City Center, which is housed in the Petronas Tower, in the north (Bunnell 2003; Indergaard 2004).

(Ramasamy, Chakrabarty, and Cheah 2004). By October 31, 2007, 1,941 companies had obtained MSC status and were operational. Application software was the principal activity for 45 percent of these companies, and mobility-embedded software and hardware was a distant second (21 percent) followed by single sign-on and multimedia. Of the companies with approved MSC status, 1,442 were Malaysian owned (51 percent ownership and higher), and 449 were foreign owned.

Initially MNCs were hesitant to invest in the MSC for a number of reasons. One was the scarcity of skilled workers. In addition, many firms had already invested in Singapore. Given uncertainty regarding demand in the region (Indonesia was the largest market in Southeast Asia but with uneven growth prospects) and having invested in Singapore, many IT companies have been slow to take advantage of incentives offered by the MSC (Fleming and Søborg 2002).

Investment in infrastructure alone is not enough. Firms need to integrate ICT into their operations, develop appropriately tailored procedures, and train a staff (Brynjolfsson and Hitt 2003). Large firms are the ones that typically take the lead in using ICT. Following their lead, small and medium-size enterprises adopt ICT, especially when there is a common platform for them to engage in business-to-business commerce. This process is off to a slow start in Malaysia. To accelerate the assimilation of ICT, the government could, through its e-Government Initiative, demonstrate the merits of incorporating ICT in improving the efficiency of company operations (World Bank 2007a). The government can also facilitate the development of common platforms so that firms can exchange data.

Access to Finance

Once innovation-led local development acquires momentum as a result of the demonstrated potential of new technologies being commercialized by pioneering firms, new start-up and spin-off activities can be boosted by the energetic midwifery of angel investors, venture capitalists, and providers of key business services. Kenney (2008) and others argue persuasively that after a few lead firms had successfully opened a new technological salient and through demonstration effects had sharpened the appetite for entrepreneurship, the availability of risk capital helped sustain innovation in Silicon Valley. Patient risk capital, mostly from public entities, also enabled first firms in Japan and later in Korea and Taiwan, China, to enter new industries, to survive in the face of years of losses, to invest in capital-intensive production facilities, to avoid the risk of takeover, and to grow into world-class companies. A similar crop of firms are now maturing in China, financed by risk capital from a variety of sources, mainly public.

Availability of Credit

Are firms in Malaysia facing difficulties in securing financing for their expansion and in investing in innovation activities? According to the World Bank's Doing Business indicators, access to finance in Malaysia is better than in other economies

Figure 6.4 M2 as a Percentage of GDP, 1995–2007

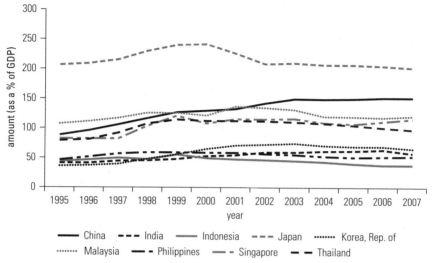

Source: World Bank World Development Indicators database.

in East Asia (annex table 2.A.6 in chapter 2; see also figure 6.4). It is Indonesia that is the laggard among the Southeast Asian countries. Although in Malaysia credit to the private sector as a share of GDP has decreased from a high of 210 percent in 1997 to 113 percent of GDP in 2006, Malaysia's ratio of credit to GDP is the second highest relative to other economies in East Asia, after Japan and China (see figure 6.5). Credit availability does not appear to be constraining business development in Malaysia, although few firms will ever pass up the opportunity to complain about the scarcity of credit and the shortage of subsidized financing.

Venture Capital

The supply of venture capital in Malaysia is still small relative to other countries in the region, although this circumstance is more a reflection of demand than supply. Interviews with staff members of Multimedia Development Corporation and Malaysia Debt Ventures suggest that the current level of financing is more than enough to fund promising new start-ups. Between 2003 and 2006, the annual commitments by venture capitalists in Malaysia had climbed by 63 percent to reach US$899 million (3 percent of fixed investment in 2006) as against US$6.0 billion in China, US$6.3 billion in Korea, US$9.8 billion in Singapore, and US$6.3 billion in Taiwan, China, in 2001 (Kenney, Han, and Tanaka 2004).

The venture capital industries in Indonesia, the Philippines, and Thailand are still at a nascent stage, and recent data are scarce. The national venture capital pool in Indonesia stood at US$153 million in 2001; only Vietnam's pool was lower. For the Philippines and Thailand, these figures were larger at US$291 million and

Figure 6.5 Domestic Credit to Private Sector, 1995–2007

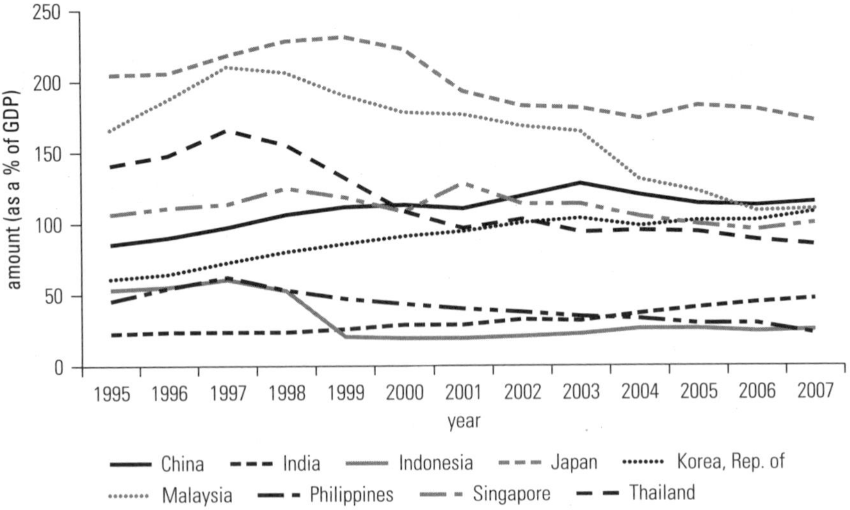

Source: World Bank World Development Indicators database.

US$580 million, respectively. The trend from 1991 to 2001 showed an initial increase followed by successive declines since 1997 for Indonesia, but continuing growth throughout the period for the Philippines and Thailand—more so for Thailand than the Philippines. The sources of funds in these countries are similar to those in other Asian economies. Corporations provided almost half of the venture capital in both Indonesia and the Philippines in 2000. In Thailand, banks (29 percent) were in the lead, followed by the corporate sector, whereas for Indonesia and the Philippines, banks were the second-largest contributor. The government was the third-largest source in Indonesia (Kenney, Han, and Tanaka 2004).[36] The prospects for further development are hampered by the lack of a suitable institutional environment, human capital, and infrastructure and, since 2008, by the global recession. Interviews with venture capitalists in East Asia reveal that the weakness of legal and other institutions leads venture capitalists to fund entrepreneurs whom they know personally. Entrepreneurs connected to government officials are favored, although too much reliance on political relationships can be a handicap once a key political figure is out of office. Furthermore, venture capitalists in all three countries face difficulties in exerting control over the firm because founders (and their family members) are frequently reluctant to relinquish control (Ahlstrom and Bruton 2006).

---

[36] Foreign venture capital investors are also a significant force in East Asia (Wright, Pruthi, and Lockett 2005).

Table 6.27 Volume of Venture Capital from Public and Private Sources in Malaysia, 2003–06

Source	Amount (US$ million)			
	2003	2004	2005	2006
Government agencies	299.5	254.1	258.6	366.2
Corporations	113.8	190.0	219.3	338.0
Foreign investors	6.1	6.0	50.6	67.5
Banks	97.3	77.1	97.4	66.0
Insurance	0.6	7.1	7.0	10.3
Pension and provident funds	0.6	0.5	12.2	6.6
Individuals	34.8	62.5	38.2	44.6
Total	552.6	597.4	683.4	899.3

Source: Economic Planning Unit of Malaysia.
Note: Exchange rates used are nominal exchange rates at the end of period averages.

To encourage the development of venture capital firms in Malaysia, the government has pursued a number of policies to attract private venture capital. It has also established public venture capital funds such as the Malaysia Venture Capital Management (MAVCAP), the Malaysian Technology Development Corporation, and the Malaysian Life Sciences Capital Fund. Alongside the financing from these public entities, private venture capital has been forthcoming from banks such as the Mayban Venture Capital Company. Currently, 49 venture capital firms operate in Malaysia. Of these, 8 are government owned, 8 are bank owned, 13 are private firms, and 6 are corporate affiliates (Lyons and Kenney 2007).[37] Malaysian government agencies account for 41 percent of the available capital, and corporations for 38 percent (see table 6.27). What is notable is that venture capital funding from corporate sources tripled between 2003 and 2008. More funding was also available from foreign venture capitalists. In contrast, venture capital funding from banks declined. The shift from reliance on government agencies and banks to corporations and foreign venture capital sources is a welcome development and indicates that the venture capital financing in Malaysia is becoming more market based.[38] Whether the supply of venture capital will continue growing given the tightening of credit and greater risk aversion following the onset of recession in 2008 and 2009 is uncertain. In the United States, venture financing has become scarce, and returns to venture capital have fallen.

---

[37] Although there are 49 VC firms identified, only 26 are registered as members of the Malaysian Venture Capital and Private Equity Association, which suggests that some of these 49 firms may be inactive (Lyons and Kenney 2007).

[38] The Israeli experience indicates that the venture capital industry has grown side by side with high-tech start-ups. For venture capital firms to be viable, the locality needs to have a large pool of potential and actual start-ups so that the firms can spread the risks (Avnimelech and Teubal 2006).

Figure 6.6 Number of Deals in Malaysia, Singapore, and Thailand, Including All Stages, 1990–2007

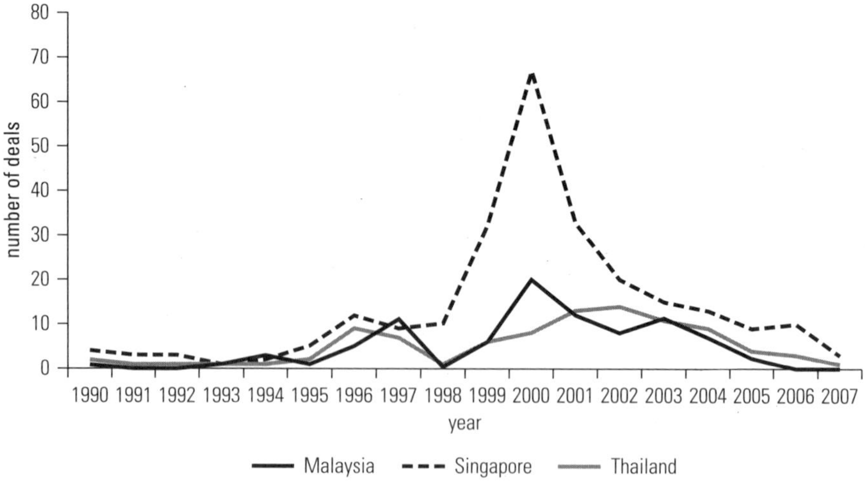

Source: Lyons and Kenney 2007.

Although the potential supply of venture capital has increased over the years, the number of deals has not. They peaked in 2000 with 20 venture capital deals and have since declined (see figure 6.6). Nonetheless, 126 firms were listed on the MESDAQ (Malaysian Exchange of Securities Dealing and Automated Trading) in late 2007 as against just 3 in 2001.[39] For start-up and early-stage financing, the peak number of deals was five in 2000 and 2001. The scarcity of deals is also apparent from the actions of venture capitalists, who have strayed far from their announced focus to fund projects in a variety of fields. For instance, MAVCAP, established in 2001, was intended for investment in technology-intensive areas. However, its actual portfolio consists of firms in outsourcing services, wood products, health food products, and other areas. Malaysian Venture, which was established in 1984 by a Singaporean venture capital firm, also intended to invest in electronics, plastics, and ceramics, but the actual investment flowed into more conventional resource-based products such as rubber and furniture (Lyons and Kenney 2007). Although the total number of deals and the distribution of such deals across various industrial sectors are hard to find, table 6.28 lists the cumulative

---

[39] Other countries also have established secondary boards specifically aimed at start-up firms to be listed. China established the SME (Small and Medium-Size Enterprise) Board at Shenzhen Exchange in 2004. Currently 126 firms are listed there, although many are in traditional industries.

Table 6.28 Cumulative Deals in Malaysia, by Sector, 1990–2007

Sector	Cumulative number of firms receiving investment by sector			
	1990	1995	2000	2007
Medical	0	1	5	9
IT Internet	0	0	14	26
IT non-Internet	0	0	7	39
Nontechnology	1	5	26	44
Total	1	6	52	118

Source: Lyons and Kenney 2007.

Table 6.29 Initial Public Offerings by Malaysian Firms

Firm	IPO date	Status	Listing
Cyber Village	August 24, 2001	Operating	Singapore Exchange
ETI Tech Corporation	March 28, 2006	Operating	MESDAQ
GPRO Technologies	June 2, 2004	Operating	MESDAQ
Media Shoppe	December 8, 2004	Delisted	MESDAQ
MyEG Services	April 13, 2005	Operating	MESDAQ
REDtone International	January 9, 2004	Operating	MESDAQ
Viztel Solutions	July 23, 2004	Operating	MESDAQ
iNix Technologies	September 13, 2004	Operating	MESDAQ
UnrealMind Interactive	June 30, 2004	Delisted[a]	MESDAQ
MEMS Technology	August 11, 2004	Operating	MESDAQ

Source: Lyons and Kenney 2007.
a. UnrealMind Interactive was taken over by Monstermob Group, a U.K. firm, in June 2005.

number of deals in Malaysia from 1990 to 2007.[40] This diffuse approach strongly suggests that deal flow is weak.

For a thriving venture capital industry, avenues to exit need to be present so that venture capital firms can realize the large capital gains. Two main exit strategies are the initial public offering (IPO) and the merger or acquisition. Table 6.29 lists the firms from which venture capital firms successfully exited. Of the 10 firms that were identified, 1 was listed in Singapore and the other 9 were listed on the MESDAQ. Subsequently two firms were delisted. Lack of IPOs in other markets, especially the NASDAQ (National Association of Securities Dealers and Automated Quotations), suggests that these firms, while successful, were not successful enough to be listed in

---

[40]The figures in table 6.28 were compiled from VentureExpert, which is based on self-reporting and Web sites of major venture capital firms in Malaysia and may not reflect their actual investments. However, they are the best data available (Lyons and Kenney 2007).

more desirable markets.[41] Hence, international investors are not likely to be looking for promising firms in Malaysia. In addition, a lack of mergers suggests that other firms did not also value Malaysian start-ups highly. This situation is in stark contrast to the experience of China, where many of the successful firms have been listed on the NASDAQ or even the New York Stock Exchange in the United States.

International experience points to two deficiencies in the Malaysian venture capital industry: the government's involvement in the provision of venture capital and the quality of universities. Direct provision of venture capital by the public entities tends to be less successful than venture capital by private providers, as is apparent from Malaysia's own experience and that of other countries.[42] China promoted the development of a domestic venture capital industry by providing public funding for investment.[43] As a result, government-owned venture capital funds account for 45 percent of the total in China. However, the public venture capital firms have not fared well. The most successful firms tend to be financed by foreign venture capital, even though these firms were started by Chinese entrepreneurs and are domestically oriented and much of the value creation occurs domestically (Zhang 2008). Korea also tried to stimulate growth by providing public seed capital, without much success. Public funding often comes with strings tied to a social agenda. Various affirmative action programs have also distorted public investment decisions. Experience in the United States suggests that when the provision of venture capital by the government has social goals rather than focusing purely on capital gains, its effect is marginal at best. Finally, significant involvement of a government-backed venture capital industry may crowd out the private industry, and the investment

---

[41] Fewness of IPOs was also notable in the United States during 2008 to 2009.

[42] Some of the problems with public venture capital can be overcome, as discussed by Lerner (2002). But given the nature of the business, stimulating the development of private domestic venture capital is preferable. In the 1990s, the Malaysian government offered tax incentives to venture capital firms. However, many venture capital firms did not take advantage of such tax incentives because of bureaucratic hurdles. Similarly, foreign investors complained that the approval process took too long for them to invest in Malaysian firms. Given that venture capital firms tend to be more time constrained than financially constrained, these bureaucratic delays were perceived to be a significant bottleneck. Furthermore, there were also other policies that had a negative impact on the development of the venture capital industry in Malaysia. For instance, foreign capital controls that were introduced after the Asian financial crisis discouraged foreign venture capitalists from investing in Malaysia.

[43] By 2007, assets under management have increased to US$28 billion from US$11.3 billion in 2003. A successful fund by the Chinese government is the National Electronic and Information Technology Development Fund (known as the *IT Fund*) with a rate of return of 86 percent over eight years.

flows may be toward those who are able to secure government subsidies rather than to those with the most promising returns.[44]

The case of Israel is illuminating. Although the government did provide some seed funding for direct investment, it also invested in domestic venture capital firms through the Yozma Program, which required these firms to raise additional funding from foreign partners, typically foreign venture capitalists. By doing so, the domestic venture capital industry was able to nurture domestic capabilities through partnering with foreign venture capital firms while developing international connectivity. China is currently trying to emulate the Yozma Program by allocating RMB 100 million for the National Venture Capital Promotion Fund to promote formation of a domestic venture capital industry. Similarly, MAVCAP offers RM 200 million through the second Outsource Partners Programme (OSP2) as a matching grant to a new venture capital fund that is managed by experienced local venture capital managers.

The quality of universities in Malaysia further dampens the deal flow. The other Southeast Asian countries confront similar problems. Without high-quality applied research at universities and major firms in areas that can be fruitful sources of innovations, the type of start-up firms seen in the United States and elsewhere will be limited in number. Finland; Ireland; Israel; Taiwan, China; and Boston and Silicon Valley in the United States—as well as, to a certain extent, India—all depend on top-ranking universities. Malaysia is still at the early stage of venture capital development. Hence, the focus of the policy should be on improving the quality of technical education in the general population as well as on improving the quality and research capacity of a few selected universities. Korean experience suggests that absent major involvement by foreign venture capital firms, a significant improvement in the quality of universities, and entrepreneurial spillovers from major domestic firms, local venture capital firms will struggle to effectively support start-up firms.

---

[44]An example is the Minority Enterprise Small Business Investment Corporation or state and local venture capital programs aimed at funneling investment into certain geographic areas of the country (Lyons and Kenney 2007).

7

From Technology Development to Innovation Capability

For Malaysia to exploit opportunities in electronics and related areas of biotechnology and medical equipment, in resource-based industries, and in processing activities, it needs to strengthen its domestic innovation capability.[1] This chapter addresses the questions that bear on the development of the knowledge economy and of a culture of innovation. Is Malaysia spending enough on research and development (R&D), and is this spending efficient? Is Malaysia focusing on the right areas? Is Malaysia reaping the benefit from R&D?[2] Are there an emerging innovation culture and a demonstrated readiness to innovate in any sector? As a foreshadowing of the findings, R&D spending in Malaysia is close to the average after controlling for the stage of development. The level of spending on R&D seems to be determined by the surprisingly modest amount of research in the electronics and the transport subsectors, both of which tend to be highly research

---

[1] Malaysia can be classified as a *passive FDI-dependent learning country* using the *crystal framework*. This means that though the industrial activities are in high-tech areas, the domestic learning capability remains weak, typically because of foreign direct investment (FDI), and the country cannot take full advantage of the FDI inflow. In contrast, Ireland and Singapore are classified as *active FDI-dependent learning countries*, complementing the inflow of FDI and high-tech activities with domestic research and development efforts (Soubbotina 2006). See also the receiver-active approach of Japanese firms described by Kodama and Suzuki (2007).

[2] This question is only partially answered in this chapter. As Michelacci (2002) and many others have pointed out, unless the stock of ideas generated by R&D is used by entrepreneurs, the returns from R&D will be suboptimal and the productivity of R&D spending will be lower. Increasing the supply of entrepreneurs willing to bet on new business models and innovations—a vast topic—is not addressed in this study.

159

Table 7.1 R&D Spending as a Share of GDP, 1996–2006

Country	Share of GDP (%)										
	1996	1997	1998	1999	2000	2001	2002	2003	2004	2005	2006
China	0.57	0.64	0.65	0.76	0.90	0.95	1.07	1.13	1.23	1.33	1.42
Indonesia	0.16	0.16	0.09	0.09	0.09	0.09	0.09	0.04	0.01	—	—
Japan	2.82	2.89	3.02	3.04	3.05	3.13	3.18	3.2	3.18	3.33	3.40
Korea, Rep. of	2.42	2.48	2.34	2.25	2.39	2.59	2.53	2.63	2.85	2.98	3.23
Malaysia	0.22	—	0.39	—	0.50	—	0.69	—	0.63	—	0.95
Singapore	1.37	1.48	1.81	1.9	1.88	2.11	2.15	2.12	2.24	2.36	2.39
Thailand	0.12	0.1	—	0.26	0.25	0.26	0.24	0.26	0.25	—	—

Sources: World Bank World Development Indicators database. The data for Indonesia are from Association of Southeast Nations Science Technology Indicator/Technology Competitiveness Indicator project (ASEAN STI/TCI) (http://aseank.kisti.re.kr/).
Note: — = not available.

intensive in more advanced economies. By the same token, the limited supply of researchers may itself reflect weak demand. Furthermore, R&D spending is spread relatively thinly across a number of technological disciplines.

R&D Spending

Knowledge intensity is measured by a number of indicators, among which R&D spending has some primacy as it affects the supply of research skills from universities, the demand for scientists and engineers, and the potential output of patents. However, R&D spending is not a guarantee of commercially viable innovation, which can help raise growth in gross domestic product (GDP). R&D spending in Malaysia as a share of GDP has increased slowly over the years to reach 0.95 percent in 2006 (see table 7.1).[3] Relative to other, more highly industrialized economies in East Asia, Malaysia's aggregate outlay on R&D is lower, although it is comparable to that of Brazil and Mexico, which have similar levels of per capita income. In fact, Malaysia was squarely on the trend line estimated for a sample of comparators for 1996 to 2004 (see figure 7.1). Not surprisingly, R&D spending by China, Japan, the Republic of Korea, and Singapore is above the predicted line, especially that by Korea.

The number of personnel engaged in R&D in Malaysia has risen at a slow pace since 1997. The number of people engaged in R&D in 2004 was 12,800 (see table 7.2), compared with, 51,500 in Indonesia (in 2001), 21,900 in Singapore, and 32,000 in Thailand. The number of researchers has tripled in Thailand in eight years while

---

[3] The Second Science and Technology Policy (STEP2), launched in 2003, aims to increase R&D spending to 1.5 percent of GDP by 2010 (Shapira and others 2006).

Figure 7.1 R&D Expenditure as a Share of GDP, 1996–2004

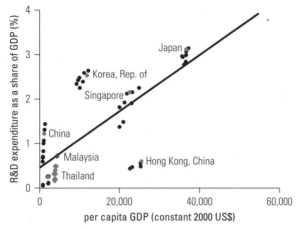

Sources: World Bank World Development Indicators database; ASEAN STI/TCI (http://aseank.kisti.re.kr/).
Note: The trend line was calculated as the predicted values from regressing R&D expenditure as a share of GDP on per capita GDP. The sample of economies includes Organisation for Economic Co-operation and Development countries plus Argentina; Brazil; Chile; China; Hong Kong, China; Indonesia; Malaysia; Mexico; the Philippines; Singapore; South Africa; Taiwan, China; and Thailand.

those in Malaysia doubled during the same period. Most are young and relatively inexperienced. Both countries lack seasoned researchers and project managers, and unlike Singapore, neither benefits much from an infusion of foreign talent. There was no change in the number of personnel engaged in R&D in the Philippines during 1998 to 2004.

Among the East Asian economies, China has the most researchers. Normalized with reference to population size, Japan has the largest number of R&D personnel,

Table 7.2 Total R&D Personnel Nationwide, 1997–2004

Country	R&D personnel (full-time work equivalent thousands)							
	1997	1998	1999	2000	2001	2002	2003	2004
China	1,427.8	1,667.7	830.0	755.0	821.7	893.0	956.5	1,035.2
Indonesia	—	—	—	56.4	51.5	—	—	—
Japan	945.8	948.1	891.8	894.0	925.6	919.1	896.8	892.1
Korea, Rep. of	156.1	152.2	135.7	136.6	128.6	137.9	138.1	89.9
Malaysia	6.7	6.7	4.4	6.7	6.7	10.1	10.1	12.8
Philippines	—	15.6	15.6	15.6	15.6	15.6	15.6	15.6
Singapore	9.5	11.1	12.1	14.9	15.1	19.4	19.5	21.9
Thailand	10.6	12.8	12.8	14.0	14.0	14.0	20.0	32.0

Source: ASEAN STI/TCI (http://aseank.kisti.re.kr/).
Note: — = not available.

Figure 7.2 Total R&D Personnel, 1999–2004

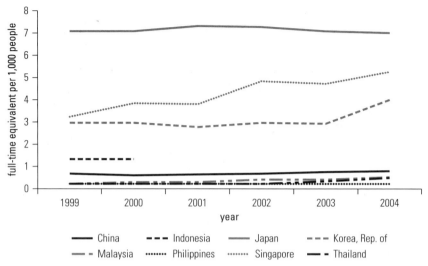

Source: ASEAN STI/TCI (http://aseank.kisti.re.kr/).

followed by Singapore and Korea, respectively. Figure 7.2 shows the researchers per 1,000 people. The ratio of researchers to the total population has increased rapidly in Singapore in the past five years. The ratios for Malaysia and Thailand, while rising, are still relatively low. And although the gap between Malaysia and other Northeast Asian economies is narrowing, it is doing so at a rate that is slower than for R&D spending (see figure 7.3). China, Japan, Korea, and Singapore have more researchers than their per capita income would indicate. Rapidly increasing the number of technical workers engaged in R&D is harder than raising spending on R&D because building up research skills takes time. However, in the absence of sufficient numbers of adequately trained researchers and experienced research project managers, the risk is that a part of the increase in R&D spending may be used inefficiently or may be devoted to downstream and relatively low-value product development.

Much of the R&D in Malaysia is performed by industry. It accounts for 65 percent of overall R&D spending in 2004[4] and includes state-owned firms such as Petronas (see figure 7.4). In 2000, the share of private firms in R&D was 58 percent, and foreign establishments accounted for 64 percent of private R&D spending (Tham 2004). This situation is fairly similar to that of other East Asian countries, except for Thailand, where the share of R&D by private industries was less than half. In other countries (excluding Indonesia, which has little R&D activity), private firms account for two-thirds to three-quarters of R&D spending.

---

[4]Figure was provided by the Economic Planning Unit of Malaysia.

Figure 7.3 Number of Researchers per Million People, 1996–2004

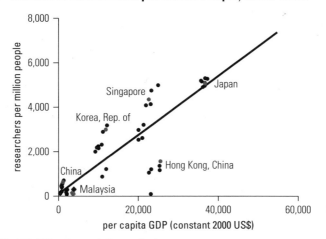

Source: World Bank World Development Indicators database.
Note: The trend line was calculated as the predicted values from regressing the number of researchers per million people on per capita GDP. The sample of economies includes Organisation for Economic Co-operation and Development countries plus Argentina; Brazil; Chile; China; Hong Kong, China; Indonesia; Malaysia; Mexico; the Philippines; Singapore; South Africa; Taiwan, China; and Thailand.

Within industry, firms in the automotive sector spend the most on R&D (RM 646.3 million), followed by firms in the office equipment and electronics subsectors (see table 7.3).[5] These three subsectors—automotive, office equipment, and electronics—account for 72 percent of the total private R&D spending (RM 2 billion). The oil and gas sector is also active in R&D, owing to the scale of the domestic industry. R&D spending by the electrical machinery subsector is less than one-third of R&D spending in the electronics industry as a whole.

R&D expenditure by the private sector in Malaysia is mainly on automotive technology, electronic components, and consumer electronics. It is relatively diffuse when compared with the distribution of R&D in other countries with similar size populations. Larger countries tend to have a more diversified portfolio of R&D spending, whereas smaller countries focus on few areas (see figure 7.5). For instance, manufacturers of pharmaceuticals conduct more than two-thirds of the R&D in Switzerland. Similarly, close to two-thirds of R&D spending in Korea is on electronics. Relative to those countries, R&D spending in Malaysia by private firms is spread more widely: 33 percent is devoted to the automotive sector, 20 percent to electronics, and 20 percent to consumer electronics.

---

[5] For a complete breakdown of private R&D spending by sectors in Malaysia, see appendix table C.1.

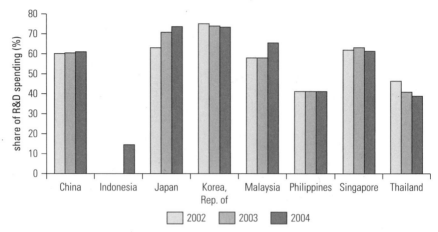

Figure 7.4 Share of R&D Spending by Businesses, 2002–04

Source: ASEAN STI/TCI (http://aseank.kisti.re.kr/).

Table 7.3 Top Five Sectors with Highest R&D Spending in Malaysia, 2006

Sector	Amount (RM million)
Manufacture of motor vehicles, trailers, and semitrailers	646.3
Manufacture of office, accounting, and computing machinery	414.4
Manufacture of radio, television, and communication equipment and apparatus	396.3
Extraction of crude oil and natural gas; service activities incidental to crude oil and natural gas extraction, excluding surveying	150.5
Manufacture of electrical machinery and apparatus, not elsewhere classified	117.8

Source: Economic Planning Unit of Malaysia.

A global R&D spending survey reveals that the automotive sector is among the most R&D intensive (Department of Trade and Industry 2007).[6] Proton Holdings Berhad is ranked 1,039th in the global top 1,250 compiled by the United Kingdom's Department of Trade and Industry, with R&D spending of £29.11 million (Department of Trade and Industry 2008).[7] Although as a share of sales (3.9 percent), R&D

---

[6] Among the global top 20 firms in R&D spending, 7 firms are in the automotive sector. Other R&D-intensive sectors are the high-tech electronics and pharmaceutical industries (Department of Trade and Industry 2008).

[7] Proton Holdings is the only firm included in the global top 1,250. Proton was ranked 629th in 2006 with spending of £53.5 million and 768th in 2007 with £39.3 million (Department of Trade and Industry 2006, 2007). R&D spending by Telekom Malaysia and Tenaga Malaysia did not qualify because it was below the cutoff line of £19 million (Department of Trade and Industry 2006).

Figure 7.5 Distribution of R&D by Sectors, 2007

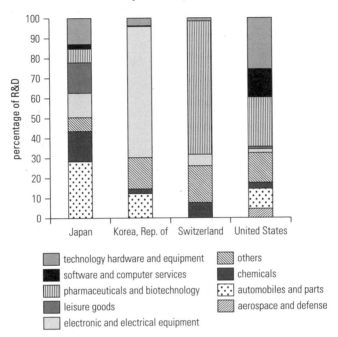

Source: Department of Trade and Industry 2007.

spending by Proton is comparable to other major automotive firms, the absolute amount of R&D spending is small relative to the major automobile assemblers, including motorcycle manufacturers (see table 7.4).

From a total of 43 government research institutes (GRIs), 40 agencies and institutes were actively engaged in R&D in 2004. Of them, the Malaysian Palm Oil Board (MPOB) was the largest spender on R&D in 2004, with an outlay of more than RM 61 million (see table 7.5). MPOB is followed by the Malaysian Institute of Microelectronic Systems (MIMOS), the Malaysian Agricultural Research and Development Institute (MARDI), and Standards and Industrial Research Institute of Malaysia (SIRIM). They are all key GRIs. Their share of the total R&D spending by GRIs was 20.9 percent, 11.4 percent, 9.0 percent, and 7.6 percent, respectively. MPOB mainly focuses on research on palm oil and its derivatives. MIMOS is oriented toward research on high-tech products.[8] MARDI is broadly responsible for research in agriculture, and SIRIM focuses on research and technology diffusion related to industrial technologies through contract research agreements with firms. Of the almost 300 researchers at SIRIM, approximately 100 work on

---

[8] The focus of MIMOS's research is on advanced informatics, microelectronics, grid computing, and cyberspace security.

Table 7.4 R&D Spending by Major Automobile Firms, 2007–08

Company	Amount (£ million)	Share of sales (%)
General Motors	4,069.13	4.4
Toyota Motor	4,005.63	3.9
Ford Motor	3,767.71	4.3
Volkswagen	3,615.87	4.5
Daimler	3,590.16	3.8
Bosch	2,614.76	7.7
Honda Motor	2,481.55	5.0
BMW	2,309.22	5.6
Nissan Motor	2,090.29	4.4
Renault	1,808.30	6.2
Peugeot	1,523.32	3.4
Tata Motors	152.43	3.4
Mahindra & Mahindra	36.92	1.6
Proton	29.11	3.9

Source: Department of Trade and Industry 2008.

advanced materials, 100 on machinery, 50 on advanced manufacturing processes, and 50 on industrial biotech and environment-related technologies. SIRIM is also the industrial standard-setting body. It is engaged in coordinating standards developed in Malaysia with other countries and is attempting to take a lead in international Halal certification in food products, cosmetics, medicines, and medical implants.

Reflecting the large R&D spending by agriculture-related GRIs (MPOB and MARDI), the top two fields of emphasis are research on crop production and research on pest and disease management. Of the top five areas of research, three are agricultural; the other two are information and communication technology (ICT) applications and civil engineering (table 7.6).

Table 7.5 R&D Spending by Top Four GRIs, 2004

Government research institute	Total R&D spending (RM million)	Share of R&D spending (%)
MPOB	61.9	20.9
MIMOS	34.0	11.4
MARDI	26.7	9.0
SIRIM	22.6	7.6
R&D spending by all GRIs	296.9	100.0

Source: Economic Planning Unit of Malaysia.

Table 7.6 Top 15 Fields of Emphasis by GRIs, 2004

Description	R&D (RM million)	Share of total (%)
Crop production	23.7	8.0
Pest and disease management	23.2	7.8
ICT applications	18.2	6.1
Agricultural engineering	15.1	5.1
Civil engineering	12.4	4.2
Horticulture	10.9	3.7
Radiation chemistry	10.5	3.5
Environmental management and bioremediation	9.3	3.1
Resource-based technology	8.0	2.7
Communication	7.4	2.5
Electrical and electronic engineering	7.3	2.5
Plant biotechnology	7.1	2.4
Process technology and engineering	6.4	2.1
Soil and water sciences	5.5	1.9
Advanced materials	5.4	1.8

Source: Economic Planning Unit of Malaysia.

Mirroring the R&D spending, MPOB and MARDI have the most researchers with 768 and 699, respectively, followed by SIRIM with 439 (see table 7.7). These research institutes are also staffed with more PhDs than other research institutes. The average number of PhDs as a share of the total staff is 25 percent for all GRIs, but more than 35 percent of researchers hold a PhD in MPOB, and close to 30 percent do in MARDI.

R&D spending by universities is concentrated in five institutions: Universiti Teknologi Malaysia (UTM), Universiti Sains Malaysia (USM), Universiti Putra Malaysia (UPM), Universiti Kebangsaan Malaysia (UKM), and University of Malaya (UM). These five universities combined account for almost 80 percent of the R&D spending of all universities (see table 7.8). Among them, the largest spender on R&D in Malaysia is Universiti Teknologi Malaysia (UTM), located in Johor (although it has a satellite campus in Kuala Lumpur). UM is located in Kuala Lumpur, and USM is in Penang. UKM and UPM are both in Selangor.

The research conducted by universities is diverse, and most fields of research absorb less than 3 percent of their total spending. The two leading fields are (a) biochemistry and (b) manufacturing and production engineering. These fields accounted for 7.2 percent and 5.8 percent, respectively, of R&D spending by universities in 2004 (see table 7.9). However, the research effort is greatly hampered by the weakness of postgraduate training and opportunities for postdoctoral research. The lack of training and research opportunities constitutes a big gap in the Malaysian R&D system that urgently needs to be closed.

Table 7.7 Distribution of R&D Researchers in GRIs with More Than 100 Researchers, 2004

Institution	Number of Malaysians					Number of foreigners					Total
	PhD	Master's	Bachelor's	Diploma	Subtotal	PhD	Master's	Bachelor's	Diploma	Subtotal	
MPOB	271	157	191	142	761	5	0	2	0	7	768
MARDI	206	280	102	110	698	1	0	0	0	1	699
SIRIM	105	119	182	33	439	0	0	0	0	0	439
FRIM	135	107	60	19	321	5	2	0	0	7	328
EPRD	46	123	41	11	221	0	0	0	0	0	221
MINT	67	46	83	6	202	3	0	1	0	4	206
MIMOS	8	34	146	2	190	2	0	6	0	8	198
DOS	2	23	128	28	181	0	0	0	0	0	181
IMR	65	46	41	17	169	2	1	1	0	4	173
Total in all institutions	1,085	1,220	1,441	535	4,281	44	8	13	1	66	4,347

Source: Economic Planning Unit of Malaysia.
Note: DOS = Department of Statistics; EPRD = Educational Planning, and Research Division, Ministry of Education; FRIM = Forest Research Institute Malaysia; IMR = Institute for Medical Research; MARDI = Malaysian Agricultural Research and Development Institute; MIMOS = Malaysian Institute of Microelectronic Systems; MINT = Malaysian Institute for Nuclear Technology Research; MPOB = Malaysian Palm Oil Board; SIRIM = Standards and Industrial Research Institute of Malaysia.

Table 7.8 R&D Spending by Top Research Universities, 2004

Institute of higher learning	R&D spending (RM million)	Share of total (%)
Universiti Teknologi Malaysia	107.2	20.9
Universiti Sains Malaysia	105.9	20.6
Universiti Putra Malaysia	79.3	15.5
Universiti Kebangsaan Malaysia	76.7	14.9
University of Malaya	41.1	8.0

Source: Economic Planning Unit of Malaysia.

Table 7.9 Top 15 Fields of Emphasis by Universities, 2004

Description	R&D spending (RM million)	Share of R&D spending (%)
Biochemistry	37.1	7.2
Manufacturing and production engineering	29.7	5.8
Communication	18.4	3.6
Education	16.5	3.2
Energy technology	13.9	2.7
Mechanical engineering	13.7	2.7
Automotive engineering	12.2	2.4
Information systems	11.6	2.3
Process technology and engineering	11.5	2.2
Electronic materials	11.0	2.1
Mathematics	10.7	2.1
Resource-based technology	9.2	1.8
Condensed matter physics	9.0	1.8
Civil engineering	8.2	1.6
Clinical medicine	8.2	1.6

Source: Economic Planning Unit of Malaysia.

Biochemistry research is dominated by USM, which is responsible for more than 90 percent of R&D spending (see table 7.10). But USM has a small research staff and is hard put to assemble a critical mass. Similarly, research on manufacturing and production engineering is concentrated at UKM. Other sector-specific research, such as research on ICT and the automotive sector, tends to be concentrated at UTM.[9] See table 7.11 for a list of the top 10 GRIs and universities in R&D spending.

---

[9] UTM still needs to ramp up its research effort and raise the content. As noted by Tan Sri Zulkifli, "Most research is done on the softer sciences like biology where you study plants and write a paper" ("Engineering Research at UTeM" 2007: 70).

Table 7.10 Research Concentration by Fields, 2004

Field	R&D spending (RM million)	Share of R&D spending (%)
Biochemistry		
Universiti Sains Malaysia	34.3	92.5
Universiti Kebangsaan Malaysia	1.5	4.0
Universiti Putra Malaysia	0.8	2.1
Manufacturing and production engineering		
Universiti Kebangsaan Malaysia	27.4	92.1
Kolej Universiti Teknikal Kebangsaan Malaysia	0.8	2.8
Universiti Teknologi MARA	0.6	2.2
Communication		
Universiti Teknologi Malaysia	7.2	39.2
University of Malaya	7.0	37.8
Multimedia University	1.7	9.3
Education		
Universiti Teknologi Malaysia	3.8	23.3
Universiti Putra Malaysia	3.0	18.4
Universiti Pendidikan Sultan Idris	2.9	17.3
Energy technology		
Universiti Teknologi Malaysia	7.4	53.1
Universiti Kebangsaan Malaysia	3.8	27.0
Universiti Sains Malaysia	1.5	10.9
Mechanical engineering		
Universiti Teknologi Malaysia	5.2	38.1
Universiti Teknologi Petronas	4.4	32.2
Universiti Kebangsaan Malaysia	1.1	8.0
Automotive engineering		
Universiti Teknologi Malaysia	7.8	63.7
Universiti Putra Malaysia	2.6	21.3
University of Malaya	1.3	10.6
Information systems		
Universiti Teknologi Malaysia	5.3	45.5
Multimedia University	1.7	14.7
Universiti Kebangsaan Malaysia	1.3	11.6
Process technology and engineering		
Universiti Teknologi Malaysia	8.0	69.9
Universiti Sains Malaysia	1.2	10.4
Universiti Putra Malaysia	0.6	5.1

Source: Economic Planning Unit of Malaysia.

As is to be expected, researchers in universities publish most of the papers that appear in professional journals. The five research universities rank at the top, followed by the GRIs (see figure 7.6). Typically, these papers report basic research or address issues of theory that are generally quite remote from practical application.

Surprisingly, the three major contributors to R&D capital (private firms, universities, and GRIs) all focus on different, apparently unrelated areas of research. R&D spending by private firms is concentrated in the automotive and electronics sectors. Research in universities is wide ranging, with an emphasis on biochemistry, biology,

From Technology Development to Innovation Capability   171

Table 7.11 Top 10 R&D Spending by GRIs and Universities, 2004

Institute	R&D spending (RM million)	Share of R&D spending (%)
Universiti Teknologi Malaysia	107.2	13.2
Universiti Sains Malaysia	105.9	13.1
Universiti Putra Malaysia	79.3	9.8
Universiti Kebangsaan Malaysia	76.7	9.5
MPOB	61.9	7.6
University of Malaya	41.1	5.1
MIMOS	34.0	4.2
MARDI	26.7	3.3
SIRIM	22.6	2.8
Universiti Teknologi MARA	20.4	2.5

Source: Economic Planning Unit of Malaysia.

Figure 7.6 Number of Published Papers in Professional Journals

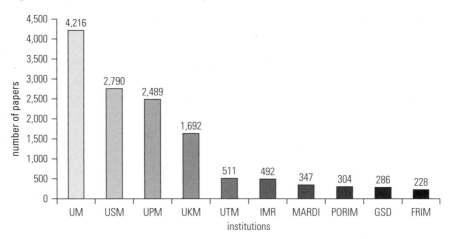

Source: Ministry of Science, Technology, and Innovation 2004.
Note: FRIM = Foreign Research Institute Malaysia; GSD = Geological Survey Department; IMR = Institute for Medical Research; MARDI = Malaysian Agricultural Research and Development Institute; PORIM = Palm Oil Research Institute of Malaysia; UKM = Universiti Kebangsaan Malaysia; UM = University of Malaya; UPM = Universiti Putra Malaysia; USM = Universiti Sains Malaysia.

and manufacturing and production engineering. GRIs devote the most attention to agricultural technologies.[10] Whether this research, measured by inputs (that is, R&D spending), is leading to innovation with commercial outcomes is considered in the next section by analyzing the patents granted in Malaysia and to Malaysian residents.

---

[10] To help develop links with industry, UTM has created the University Industry Center to foster ties and promote consulting.

Patenting Activity

Since 2001, patents granted in Malaysia have increased, with the number more than doubling between 2005 and 2006. The majority of these patents are to foreign residents (see table 7.12), although the shares of residents have almost doubled from 2006. But because the surge in patenting activities by domestic entities is recent, one cannot say whether it is the start of a trend; however, the number of patents granted to local entities continued growing through 2007.

Between 1989 and 2006, the Malaysian Patent Office granted 281 patents to private firms. There is no single dominant patent holder among the private firms. Most have only one patent. Of those firms with multiple patents, many are multinational corporations (MNCs), such as Sony EMCS and Motorola (see table 7.13).

The list of the top 10 patent owners from 1995 to 2007 is dominated by GRIs, universities, and MNCs (see table 7.14). MPOB was granted the largest number of patents during this period (22), followed by Sony EMCS (12). UPM leads among the universities with 11 patents, followed by USM (10) and UTM (8). Among the MNCs, Sony EMCS, Matsushita, and Motorola are in the top 10. Patents issued to Malaysian residents are mainly for operational technology, chemistry and metallurgy, human necessities, physics, and electricity (see table 7.15).

Patenting by entities in Malaysia with the U.S. Patent and Trademark Office (USPTO) is low, relative to other East Asian economies (see table 7.16). A total of 158 patents were granted to Malaysia-based entities in 2007. Although this number is four times more than the patents granted to residents in Indonesia, the Philippines, and Thailand combined, it is just around 40 percent of patents granted to

Table 7.12 Patents Granted in Malaysia, 1995–2007

Year	Number of patents			Share of foreign patents (%)
	Foreign	Local	Total	
1995	1,719	30	1,749	98.3
1996	1,721	81	1,802	95.5
1997	732	55	787	93.0
1998	543	22	565	96.1
1999	682	38	720	94.7
2000	384	21	405	94.8
2001	1,451	19	1,470	98.7
2002	1,456	32	1,488	97.8
2003	1,549	31	1,580	98.0
2004	2,321	25	2,346	98.9
2005	2,473	39	2,512	98.4
2006	6,560	187	6,747	97.2
2007	6,645	338	6,983	95.2

Source: Intellectual Property Corporation of Malaysia (http://www.mipc.gov.my/images/stories/AR/2007/ar2007.pdf).

Table 7.13 Top Holders of Malaysian Patents: Private Firms, 1989–2006

Owner	Number of patents
Sony EMCS (Malaysia)	7
Motorola Semiconductor	5
Carsem (M)	3
EPS Systems & Engineering Sdn Bhd	3
Lysaght (Malaysia)	3
Sunflower Industries	3

Source: Economic Planning Unit of Malaysia.

Table 7.14 Top 10 Patent Owners in Malaysia, 1995–2007

Owner	Number of patents					Total, 1995–2007
	1995	2000	2005	2006	2007	
MPOB	1	0	2	5	8	22
Sony EMCS (Malaysia)	0	0	0	7	5	12
Universiti Putra Malaysia	0	0	0	4	6	11
Universiti Sains Malaysia	1	0	0	1	3	10
SIRIM Berhad	0	0	1	5	1	8
Universiti Teknologi Malaysia	0	0	0	2	5	8
Forest Research Institute Malaysia	0	0	0	0	0	6
Patrick Cyril Augustin	0	2	0	0	0	6
Telekom Malaysia Berhad	0	0	1	3	2	6
Matsushita Home Appliance R&D Centre (M)	0	0	0	0	5	5
Motorola Semiconductor	0	0	0	3	0	5
Board of the Rubber Research Institute of Malaysia	2	0	0	1	0	5
Silterra Malaysia	0	0	0	0	5	5

Source: Intellectual Property Corporation of Malaysia.

residents of Singapore. However, patents granted to Malaysian recipients are on the rise, almost tripling since 2000, albeit starting from a low base.

Relative to other Asian economies and higher-income countries, U.S. patents per 1,000 people in Malaysia are lower but close to what the per capita income predicts (see figure 7.7). Patenting by Japan; Korea; Singapore; and Taiwan, China, is higher than the average adjusted for their per capita incomes. Hong Kong, China, which spends little on R&D, still manages to receive U.S. patents comparable to its income level.

A closer look at the patents granted to Malaysian organizations by the USPTO reveals that the bulk of the U.S. patents are granted to MNCs located in Malaysia

Table 7.15 Distribution of Domestic Patents by Technology Class, 2000–07

	Number of patents							
Class	2000	2001	2002	2003	2004	2005	2006	2007
Human necessities	61	155	206	224	325	333	948	1,179
Performance of operations, transport	59	233	236	242	377	452	1,155	1,213
Chemistry, metallurgy	110	288	334	396	625	600	1,275	1,748
Textiles, paper	8	18	19	28	25	30	101	109
Fixed constructions	19	44	42	38	50	82	197	221
Mechanical engineering, lighting, heating, weapons	42	102	104	119	132	164	448	408
Physics	36	231	228	190	321	316	1,042	883
Electricity	70	399	323	341	492	531	1,583	1,222
Total	405	1,470	1,492	1,578	2,347	2,508	6,749	6,983

Source: Intellectual Property Corporation of Malaysia.

Table 7.16 Patents Granted by U.S. Patent and Trademark Office to Foreign Residents

	Number of patents							
Economy	2000	2001	2002	2003	2004	2005	2006	2007
Korea, Rep. of	3,314	3,538	3,786	3,944	4,428	4,352	5,908	6,295
Taiwan, China	4,667	5,371	5,431	5,298	5,938	5,118	6,361	6,128
China	119	195	289	297	404	402	661	772
Singapore	218	296	410	427	449	346	412	393
Hong Kong, China	179	237	233	276	311	283	308	338
Malaysia	42	39	55	50	80	88	113	158
Philippines	2	12	14	22	21	18	35	20
Thailand	15	24	44	25	18	16	31	11
Indonesia	6	4	7	9	4	10	3	5

Source: USPTO.
Note: Economies are ordered by the number of patents granted in 2007.

(see table 7.17). Apart from the individually owned patents, only four Malaysian organizations—Silterra, MPOB, Harn Marketing, and UPM—were granted five or more patents each between 2003 and 2007.[11]

Table 7.18 lists the top patent technology classes granted to Malaysian residents from 2003 to 2007. The largest patent classes are associated with the semiconductor

---

[11] The USPTO lists an organization only if it has at least five patents granted during the period.

Figure 7.7 Malaysia's U.S. Patents Relative to Those of Other Countries, 1977–2006

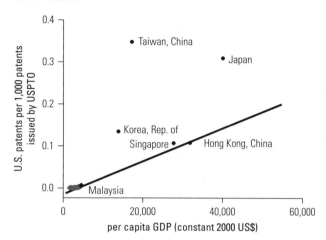

Source: USPTO.
Note: The trend line was calculated as the predicted values from regressing the number of patents per 1,000 patents issued by the USPTO on per capita GDP. The sample of economies includes Organisation for Economic Co-operation and Development countries plus Argentina; Brazil; Chile; China; Hong Kong, China; Indonesia; Malaysia; Mexico; the Philippines; Singapore; South Africa; Taiwan, China; and Thailand.

industry, both for devices and manufacturing processes, reflecting the importance of the semiconductor industry in Malaysia. Other technology classes are also mainly in the electronics industry, although some are related to the medical device sector. This sector is potentially a growth area, but much will depend on the readiness of electronics firms to diversify into bioinformatics and bioassays and electronic implants.

Many economies are engaged in semiconductor manufacturing activities. The number of patents granted by USPTO to these economies far exceeds those in Malaysia (see table 7.19). In the five years from 2003 to 2007, 52 patents were granted to residents in Malaysia in the active solid-state devices technology class. This number is small compared with others—for example, 7,923 for Japan; 1,843 for Taiwan, China; and 1,507 for Korea—and it reflects the technological gap between these countries and Malaysia even though Malaysia's manufacturing capability in the area of electronics is second to none.

The pattern of patenting in the manufacturing process technology class is similar. Although it represents the second most active patenting technology area by Malaysian residents, the 45 patents in this area are dwarfed by the numbers of patents granted to residents in other economies (see table 7.20). So far, Malaysia is ahead of China, Israel, and the Philippines. However, patenting by residents in China has been increasing in recent years, and Chinese residents received more patents overall in 2007 than did residents of Malaysia (see table 7.16). Thus, the head start that Malaysia has enjoyed may be quickly narrowing.

Table 7.17 U.S. Patents Granted to Malaysian Organizations, 2003–07

First-named assignee	Number of patents					
	2003	2004	2005	2006	2007	Total
Avago Technologies Ecbu IP (Singapore)	0	0	0	12	50	62
Intel	5	4	11	11	27	58
Individually owned patent	6	13	6	25	6	56
Agilent Technologies	2	11	22	3	1	39
Altera	1	1	2	7	13	24
National Semiconductor	3	7	5	3	4	22
CeramOptec Industries	6	3	5	4	0	18
Freescale Semiconductor	0	0	4	1	8	13
Motorola	2	4	1	4	2	13
Avago Technologies General IP (Singapore)	0	0	0	4	8	12
Advanced Micro Devices	3	1	1	3	1	9
Grossman Product Services	1	5	0	1	0	7
Semiconductor Components Industries	1	0	2	3	1	7
Chartered Semiconductor Manufacturing	3	2	1	0	0	6
Malaysian Palm Oil Board	1	0	0	3	2	6
Silterra Malaysia	0	3	1	1	1	6
Texas Instruments	0	0	2	2	2	6
Harn Marketing	0	0	1	2	2	5
Serac Group	0	3	2	0	0	5
Universiti Putra Malaysia	1	1	2	1	0	5

Source: USPTO.

It appears that indigenous Malaysian firms do very little patenting. More than 90 percent of patents granted by the Malaysia Patent Office are to foreign residents. Only a few are granted to local residents, the majority of which are to MNCs. Except for patents granted to individuals, only a small number of recent U.S. patents were granted to MNCs located in Malaysia. Patents granted by the USPTO to Malaysian residents are mainly those used in the semiconductor and electronics industries. To a certain extent, this finding matches with the R&D spending by private firms in Malaysia, although it seems that the USPTO has granted no patents related to automotive technologies, engineering, biotechnology, or agro-processing technology to Malaysian residents.[12]

In sum, R&D spending in Malaysia is low relative to other economies in East Asia. Moreover, the patents granted to indigenous Malaysian organizations are

---

[12]Pinheiro-Machado and Oliveira (2004) note that academic patenting in Brazil is constrained by the lack of specialized staff members to assist with the process and the cost associated with filing. Others have emphasized the cost factor as well.

Table 7.18 U.S. Patent Classes Granted to Malaysian Residents, 2003–07

Class	Number of patents					
	2003	2004	2005	2006	2007	Total
Active solid-state devices (for example, transistors, solid-state diodes)	3	9	12	11	17	52
Semiconductor device manufacturing, process	5	7	8	9	16	45
Radiant energy	1	1	1	14	18	35
Illumination	1	0	2	5	13	21
Electricity, measuring and testing	1	3	6	4	6	20
Plastic and nonmetallic article shaping or treating, processes	0	5	0	4	1	10
Electronic digital logic circuitry	0	0	0	3	7	10
Electric lamp and discharge devices	0	2	0	4	3	9
Electric lamp and discharge devices, systems	0	1	2	4	2	9
Computer graphics processing and selective visual display systems	0	1	1	2	5	9
Electricity, electrical systems and devices	0	1	4	2	2	9
Metal working	2	2	0	3	1	8
Data processing, design and analysis of circuit or semiconductor mask	0	0	1	3	4	8

Source: USPTO.

Table 7.19 Number of U.S. Patents Granted in the Active Solid-State Devices Technology Class, by Geographic Origin, 2003–07

Economy	Number of patents					
	2003	2004	2005	2006	2007	Total
United States	1,717	1,819	1,688	1,699	1,536	8,459
Japan	1,573	1,798	1,519	1,575	1,458	7,923
Taiwan, China	303	353	357	415	415	1,843
Korea, Rep. of	244	246	259	345	413	1,507
Singapore	34	53	37	50	56	230
Malaysia	3	9	12	11	17	52
Israel	12	7	10	4	9	42
China	5	7	10	3	8	33
Philippines	7	5	5	6	5	28
Thailand	1	0	0	1	0	2

Source: USPTO.
Note: Economies are ordered by the number of patents granted in 2007.

Table 7.20 Number of U.S. Patents Granted in Semiconductor Device Manufacturing, Process Technology Class, by Geographic Origin, 2003–07

Economy	Number of patents					Total
	2003	2004	2005	2006	2007	
United States	2,321	2,299	2,014	2,068	2,011	10,713
Japan	1,228	1,240	1,090	1,147	1,009	5,714
Korea, Rep. of	456	504	466	595	614	2,635
Taiwan	607	610	447	490	421	2,575
Singapore	107	73	65	68	57	370
Malaysia	5	7	8	9	16	45
China	2	3	9	12	14	40
Israel	13	7	2	9	6	37
Philippines	1	6	5	4	3	19

Source: USPTO.
Note: Economies are ordered by the number of patents granted in 2007.

few, suggesting that innovation capability, as measured by patents received by indigenous Malaysian entities, is very limited. This limited level of innovation is comparable to that found in Indonesia and Thailand, for example, but it is far behind the level of Taiwan, China, and the Northeast Asian economies.

Licensing and Technology Transfer

Data on royalty and technology licensing fees can reveal the flow of technology trade among countries and can indicate whether a country is a net technology creator or user. Measured by receipts, Korea is by far the largest recipient of royalties and licensing fees in East Asia (excluding Japan). The acceleration in receipts started in 1988. Royalty and licensing fee receipts in Singapore are also on an upward trend (see figure 7.8). This finding suggests that both Korea and Singapore have been developing technological capabilities and exporting technology in recent years, especially Korea.[13]

Technology absorption is reflected in the rising royalties and licensing fees paid by China and Singapore. The growth in China has been tremendous, a more than 10-fold increase since 1998. Payments by other countries, including Malaysia, also increased, albeit at a much slower pace (see figure 7.9).

In East Asia, Japan is the only country that has a positive balance of payments in royalty and licensing fees. It became a net provider of technologies only after 2003.

---

[13] U.S. patents granted to the Korean company Samsung also increased rapidly during this period. In 1997, Samsung ranked 17th in total number of patents granted (587 patents) in that year. By 2006, Samsung ranked second, with 2,451 patents.

Figure 7.8 Royalty and License Fee Receipts, 1995–2006

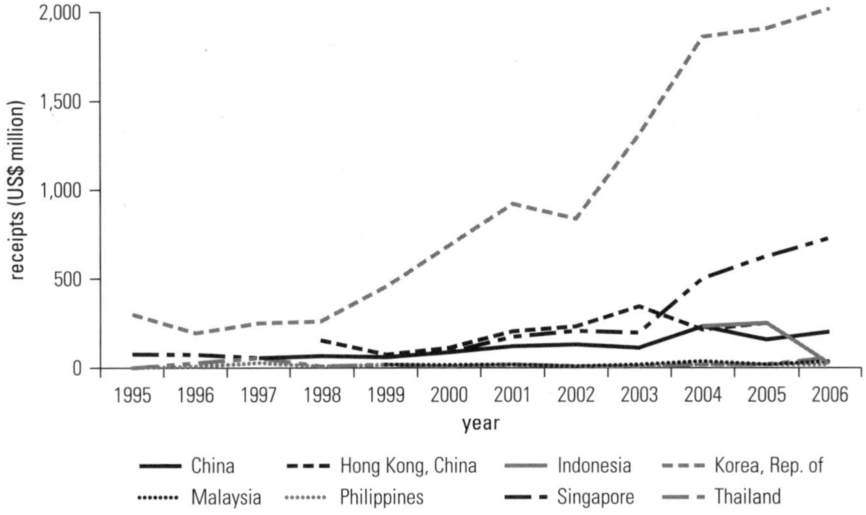

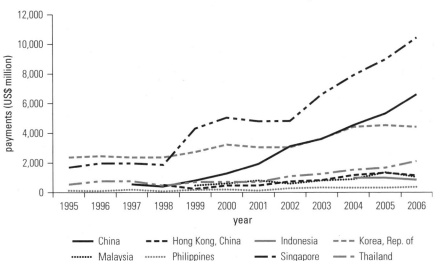

Source: World Bank World Development Indicators database.

Until then, the balance of payments was negative. Similarly, other countries are mainly users of technologies developed elsewhere. Relative to China and Singapore, net payments by Thailand are inching up gradually (see figure 7.10). This finding implies that both China and Singapore are actively licensing technologies developed

Figure 7.9 Royalty and License Fee Payments, 1995–2006

Source: World Bank World Development Indicators database.

Figure 7.10 Net Royalty and Licensing Payments, 1995–2006

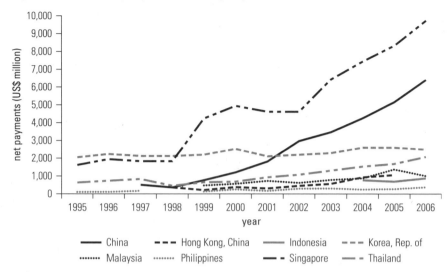

Source: World Bank World Development Indicators database.

elsewhere to be incorporated in their products. It could reflect an acceleration of their rate of catch-up (which is apparent in the case of China) and the faster incorporation of technology content in their exports. Korea's licensing payments and receipts are increasing in tandem, similar to the experience of Japan.

In countries where royalty and licensing fees are increasing less, the product mix (or the technological content of products) may be changing more slowly, given the low levels of R&D expenditure and of patenting activities (both domestically and in the United States). Malaysia and the neighboring Southeast Asian countries are all in this category, although Singapore is on a steady upward trend and, since 2003, Malaysia's licensing payments have also started rising.

Research Activities of Malaysian Firms

The private business sector accounts for a large fraction of R&D spending in Malaysia, and much of the R&D activities tend to focus on incremental innovation and process improvements.[14] This is especially true for the R&D centers of MNCs. Most of the key R&D is conducted at headquarters locations. Local subsidiaries, even if they have an R&D department, tend to concentrate on downstream product development and adaptation for local requirements or production process improvements. For instance, Advanced Micro Devices (AMD) employs 30 workers who are primarily occupied with process innovation rather than contributing to

---

[14] A survey of 1,800 firms in Malaysia by Shapira and others (2006) also shows that manufacturing firms in Malaysia are more adept at adopting and adapting technologies than at generating new technology.

the design of AMD's chips, although 80 percent of AMD's chips are produced in Penang.[15] And DENSO, with a small R&D facility in Malaysia, employs 18 workers, who are mainly engaged in adapting technology for local requirements. Motorola might be an exception because it has largely localized its two-way radio development and production in Penang, and it has close to 800 R&D employees who are engaged in this effort. Agilent Technologies is also concentrating more of its production of testing and measuring equipment in Penang.

Domestic firms mainly undertake incremental product and process innovations. Tai Kwong Yokohama, a battery maker for the automotive sector, does not have a formal R&D center. Its R&D activities are based on individual projects done in house. Because automotive batteries are a standardized item with specific predetermined dimensions, much of the R&D relates to process innovation rather than product innovation. An innovative solution with both environmental and cost implications is a lead reclamation plant set up by the firm to recover lead from used batteries (Wong, Chang, and Cheng 2007). Petronas spends only 1.5 to 2.0 percent of its revenue on R&D, devoting much of it to improving process engineering and achieving operational efficiencies.

Even so, some firms are starting to take on product innovation. For instance, Straits Orthopaedics is aiming to develop a "system" (a complete suite of orthopedic items) rather than components by spending close to 15 percent of its revenues on R&D. Straits Orthopaedics was established in 2003 as a joint venture with a U.S. partner. The impetus for establishing this firm came from the slowdown in the automation and semiconductor industries. The automation industry in Penang is closely linked with the semiconductor industry. When the growth of the semiconductor industry slowed as MNCs shifted their investments to China, the local opportunities for continuing in the machinery industry seemed less promising. The choice facing the founder of Straits was either to follow MNCs into China or to seek new business opportunities. Globally, only five or six large firms control 80 percent of orthopedic products. With the impending reform of the health care sector in Japan (a mandatory 20 percent reduction in medical devices, including orthopedics, to rein in the rising health costs), the founder saw an opportunity to enter this market by taking advantage of the advanced manufacturing capabilities that existed locally together with the relatively lower wages. The manufacturing skills acquired through the production of semiconductors are an important asset, because the material and production quality control requirements for orthopedics are much higher than for other manufactured goods. In addition, the manufacturing facilities need to be registered with and approved by the regulatory bodies on health of importing countries (in the case of the United States, the Food and Drug Administration). Currently, Straits Orthopaedics serves as the subcontractor for

---

[15]The future of this production facility is uncertain in view of AMD's decision in October 2008 to spin off its chip production operations (Global Foundries) into a separate company, jointly owned with Advanced Technology Investment from Abu Dhabi, and to concentrate on chip design.

major orthopedic manufacturers. Blueprints for many products are given to the firm. Straits Orthopaedics makes its own machines. However, raw materials are all sourced from the United States and are shipped by air because time is critical. Some of the raw materials are becoming harder to secure because of rising demand from China. Of 330 workers, 150 are engineers (of which 60 are foreign engineers). The firm recruits from a global pool of fresh graduates, although the firm recently started to recruit more from local universities.

Pentamaster is another leading Malaysian manufacturer. Pentamaster, a factory automation firm, initially catered to the semiconductor industry but now offers generalized solutions to industries such as food and automotive products. Close to 80 percent of its products are exported. At the beginning, it concentrated on customizing imported machinery, but now it offers both complete hardware and complete software solutions to factory automation, although some components are still imported. The use of radio-frequency identification has increased the speed of automation tasks.

Pentamaster spends 10 percent of its revenue on R&D. It has 40 workers dedicated to R&D, although up to 200 workers of a total of 600 (350 are engineers and 200 have programming skills) may be engaged in R&D-related activities (Wong, Chang, and Cheng 2007). R&D activities are focused on two areas: (a) keeping pace with the rapidly changing technology within this sector and (b) expanding Pentamaster's operations to other industrial sectors and gaining the ability to offer turnkey solutions to other manufacturers. Traditionally, factory automation systems depended on the customization done by manufacturers. The factory automation firm provides customized machinery to fit the specifications provided by the manufacturers. Pentamaster offers a complete system (with customization done by Pentamaster) to the manufacturer by leveraging its knowledge of modular design of machinery. This capability allows buyers to reuse the module for each component (Wong, Chang, and Cheng 2007). An example of the innovative products offered by the firm is its visual inspection software to detect defects in products automatically. This innovation makes a valuable contribution to the productivity of semiconductor manufacturing.

Hovid, a pharmaceutical firm, has a number of new products arising from its own R&D, such as natural vitamin E supplements extracted from palm oil. It has two R&D facilities. The facility located at USM does fundamental research associated with drugs and health products. The other facility, located at the factory, mainly assists with process innovation. A total of 40 researchers work in R&D activities, of which 30 researchers are located at the factory. Annual R&D spending averages about 4 percent of revenues (Wong, Chang, and Cheng 2007).

Although some Malaysian firms emphasize R&D and improve technological capabilities, they are still a small minority.[16] Penang is an attractive location for

---

[16] A study by Ariffin and Figueiredo (2004) finds that 51 out of 58 sampled electronics firms are at the stage of intermediate innovation capabilities, focusing on process and incremental innovation, with some engaging in full automation, designs, and prototyping activities.

Pentamaster because of the concentration of key manufacturers (customers) and the availability of workers who understand modern manufacturing processes. At the same time, the agglomeration of these firms in Penang makes hiring workers more difficult, especially the presence of MNCs for which many workers prefer to work. In addition, the quality of education by local institutes seems to be declining, as is the motivation of the younger generations. Although Pentamaster recruits internationally, it faces difficulties because Malaysia is typically not a top destination for the best foreign workers and acquiring high-level skilled workers from other countries is difficult. Another problem is that there are only a few second-tier suppliers in Penang, forcing the firm to import a substantial number of parts. The bulk of products that Pentamaster produces are destined for foreign markets, mainly aimed at second-tier manufacturers that also need to have good-quality manufacturing capabilities.

Innovation Comparative Advantage

Few Malaysian firms view applied research as integral to their longer-term competition strategy. The upshot of this approach is most apparent in the persistent inability of firms in high-tech industries, such as electronics and auto parts, to move up the value chain. Innovation is focused on midlevel and low-tech activities that involve little formal R&D. The orientation of the innovation activities is captured by the indexes measuring "innovation comparative advantage." The estimation technique is similar to that for the revealed comparative advantage index described in chapter 3.[17] The results are presented in tables 7.21, 7.22, and 7.23 for 1995, 2000, and 2007, respectively, for the top 10 items with the greatest

---

[17]The index of innovation revealed comparative advantage is a measure of the country's innovative capacity. It is a ratio of sector i's share in country j's total patents to sector i's share in the world:

$$IRCA = \frac{Patents_{ij} / \sum_i Patents_{ij}}{\sum_j Patents_{ij} / \sum_i \sum_j Patents_{ij}},$$

where i denotes the sector, j is the country, $i = 1, 2 \ldots I$, and $j = 1, 2, \ldots J$.

Alternatively, it can also be expressed as a ratio of country j's share of patents in sector i, to country j's share of total patents (across I sectors) in the world (across I sectors and J countries):

$$IRCA = \frac{Patents_{ij} / \sum_j Patents_{ij}}{\sum_i Patents_{ij} / \sum_i \sum_j Patents_{ij}}.$$

For a given sector i, a value of this index greater than one indicates that the country j has a comparative advantage in innovation in that sector.

Table 7.21 Index of Innovation Revealed Comparative Advantage, by Technology Class: Malaysia, 1995

Class	Class title	Index value
108	Horizontally supported planar surfaces	105.8
215	Bottles and jars	93.5
33	Geometrical instruments	45.4
198	Conveyors, power driven	44.4
362	Illumination	26.9
455	Telecommunications	24.9
428	Stock material or miscellaneous articles	7.4

Source: USPTO.

innovation advantage. There is little evolution between 1995 and 2000: conveyors, glass bottles, and geometrical instruments, for example, were the focuses of innovation in both 1995 and 2000. Table 7.23 suggests that innovation advantage is diversifying, with metallurgy, auto parts, and chemical compounds entering the list, possibly reflecting academic research and research by the auto sector and petrochemical firms.

By comparison, Singapore's innovation revealed comparative advantage in the electronics area has strengthened. In 1995, none of the top 10 areas of innovation was related to electronics. By 2000, three mid-tech items had entered the list of the top 10. Seven years later in 2007, four high-tech electronics products had emerged as areas of comparative innovation advantage, suggesting that Singapore's efforts to raise R&D were beginning to pay off (see table 7.24).

Table 7.22 Index of Innovation Revealed Comparative Advantage, Top 10 Technology Classes: Malaysia, 2000

Class	Class title	Index value
406	Conveyors, fluid current	306.1
249	Static molds	107.1
65	Glass manufacturing	21.4
411	Expanded, threaded, driven, headed, tool-deformed, or locked-threaded fastener	18.8
241	Solid material comminution or disintegration	17.8
216	Etching a substrate, processes	17.4
126	Stoves and furnaces	16.8
585	Chemistry of hydrocarbon compounds	16.2
331	Oscillators	14.0
33	Geometrical instruments	11.9

Source: USPTO.

Table 7.23 Index of Innovation Revealed Comparative Advantage, Top 10 Technology Classes: Malaysia, 2007

Class	Class title	IRCA
126	Stoves and furnaces	39.4
419	Powder metallurgy processes	36.6
249	Static molds	27.0
312	Supports, cabinet structure	24.0
301	Land vehicles, wheels and axles	17.1
65	Glass manufacturing	11.1
315	Electric lamp and discharge devices, systems	10.7
585	Chemistry of hydrocarbon compounds	9.6
264	Plastic and nonmetallic article shaping or treating, processes	9.5
330	Amplifiers	9.4

Source: USPTO.

All the indicators examined in this section convey a broadly similar message: as yet, R&D has acquired limited traction in Malaysia's business sector and in academia. Research by the GRIs is dispersed over a number of fields and has not created innovation capacity in electronics and electrical engineering, auto parts, biofuels, food processing, wood products, or petrochemicals, all of which are among the principal industrial subsectors for Malaysia. Where it exists, innovation advantage is typically in low- and medium-technology areas. This finding jibes with earlier findings in this study on revealed comparative advantage, on fast-growing exports, and on new starts.

Table 7.24 Index of Innovation Comparative Advantage, Top 10 Technology Classes: Singapore, 2007

Class	Class title	IRCA
228	Metal fusion bonding	20.6
105	Railway rolling stock	9.3
232	Deposit and collection receptacles	7.7
271	Sheet feeding or delivering	7.4
451	Abrading	7.0
330	Amplifiers	5.5
294	Handling, hand and hoist-line implements	5.4
257	Active solid-state devices (for example, transistors, solid-state diodes)	5.2
708	Arithmetic processing and calculating (electrical computers)	5.1
438	Semiconductor device manufacturing, process	5.1

Source: USPTO.

Although Malaysian policy makers are gazing at the technological peaks, the reality is much more mundane. Malaysian technological capability is modest, and the trend over the past decade is not indicative of much improvement. R&D in low- and medium-tech resource-intensive industries could yield larger returns than R&D in the more glamorous high-tech activities. For example, Malaysian researchers and firms might tackle the lower slopes of the electronics sector and the petrochemical sector first. Once the capability to introduce incremental product and process innovation is firmly rooted, domestic firms would be ready for steeper areas. Other Southeast Asian countries are in similar predicaments: wanting to upgrade but lacking the innovation capability and needing to first cut their teeth on low-tech products and services before attempting to upgrade higher-tech items.

8

Can the Tigers Grow Fast and Furious Again?

Long-Run Growth

The analysis and indicators presented earlier suggest why Malaysia's growth and that of other Southeast Asian countries have slowed. The worldwide recession that commenced in 2008 and resulted in declining exports from Southeast Asian countries has further raised the odds against a return to rates of growth in excess of 6 percent per year. Nevertheless, rates in this range may be feasible for Malaysia over the medium term, assuming a robust revival of world trade. This chapter provides a perspective on the growth potential of the Malaysian economy and of some comparators by tracking the growth in per capita gross domestic product (GDP) over long time spans and by showing how countries have achieved trend accelerations by increasing the expenditure on research and development (R&D) and expanding the enrollment of tertiary-level students.

The growth rate of major Organisation for Economic Co-operation and Development (OECD) countries has been notably stable (figures 8.1–8.3). The long-run growth rate of per capita GDP for the United States between 1860 and 2000 was remarkably steady at about 1.9 percent (Jones 2005).[1] The stability of long-run growth over a comparable period is apparent for European countries, such as Germany and the United Kingdom, despite their involvement in major wars. Furthermore, the growth rate of Germany, at 1.8 percent, is similar to that of the United States. The United Kingdom has maintained a long-term trend rate of 1.4 percent. Since 1960, there has been an apparent acceleration in the growth rate of these three OECD countries to nearly 2.2 percent—a small increase that is plausibly related to the sharp increase in outlays on elements of national innovation systems (see table 8.1). Absent this infusion of resources into technology

---

[1] Jones's (2005) critique maintains that long-term growth seems to be impervious to short-term policies for stimulating performance.

Figure 8.1 Per Capita GDP Growth of Germany, 1860–2003

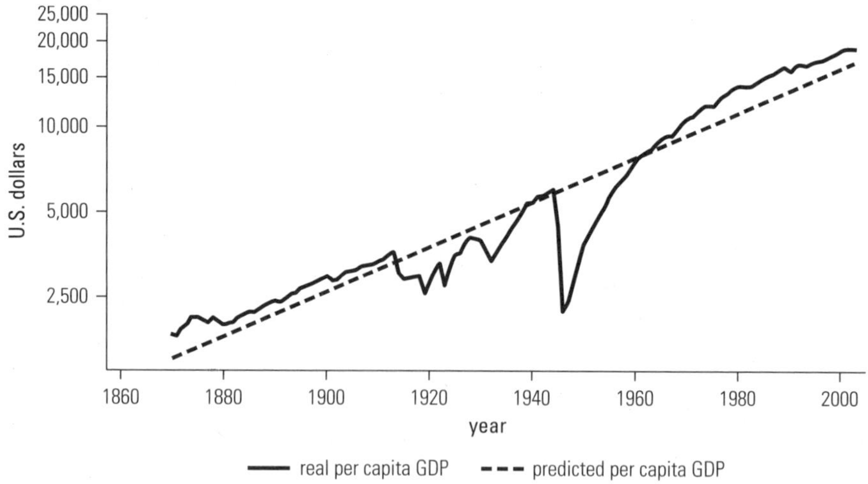

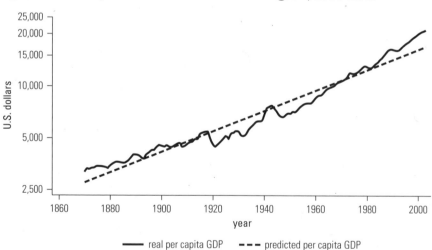

Source: Maddison 2006.

Figure 8.2 Per Capita GDP Growth of the United Kingdom, 1860–2003

Source: Maddison 2006.

development, it is possible that not only would countries have been unable to raise their per capita growth rates, but they might also have dropped below their earlier trend growth rates. This analysis argues for investment in the knowledge economy, but it does not tell us how much countries should spend, and it gives very little guidance on how much acceleration in growth rates is likely. The recent

Figure 8.3 Per Capita GDP Growth of the United States, 1860–2003

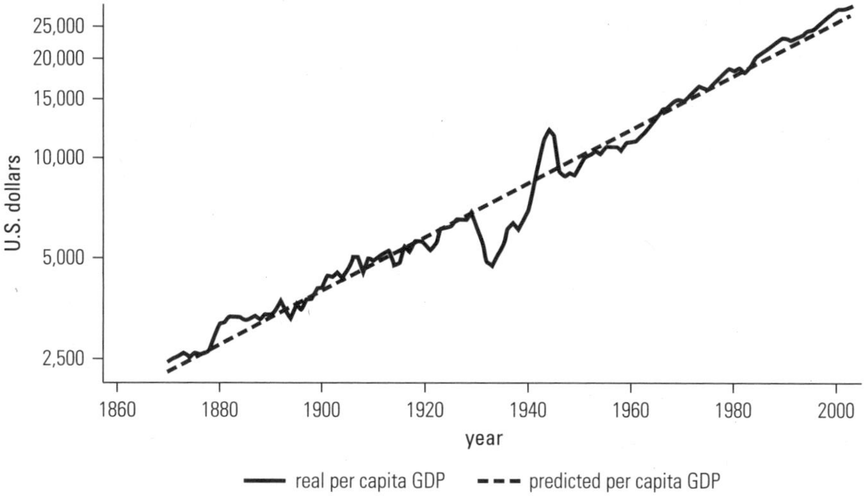

— real per capita GDP   --- predicted per capita GDP

Source: Maddison 2006.

Table 8.1 Average Growth Rates of GDP Per Capita, 1870–2003 and 1960–2003

	Average growth rate (%)	
Economy	1870–2003	1960–2003
OECD comparators		
Germany	1.8	2.2
United Kingdom	1.4	2.2
United States	1.9	2.2
East Asian economies		
Indonesia (1949–)	2.8	2.9
Japan	2.5	4.1
Korea, Rep. of	3.2	5.8
Malaysia	3.7	4.0
Philippines	2.4	1.2
Taiwan, China	3.5	5.6
Thailand	4.1	4.5

Source: Maddison 2006.
Note: Data in the first column are for 1949–2003 for Indonesia; 1911–2003 for the Republic of Korea; 1947–2003 for Malaysia; 1946–2003 for the Philippines; 1912–2003 for Taiwan, China; and 1950–2003 for Thailand.

experience of middle- and high-income countries shows that the return on R&D rises as countries approach the technological frontier (Aghion and Howitt 2005). However, the experience of Japan—and to a lesser extent the Republic of Korea— also suggests that there can be declining returns from an outlay on knowledge-deepening activities once spending on R&D begins to approach 3 percent of GDP.

Figure 8.4 Per Capita GDP Growth of Indonesia, 1960–2003

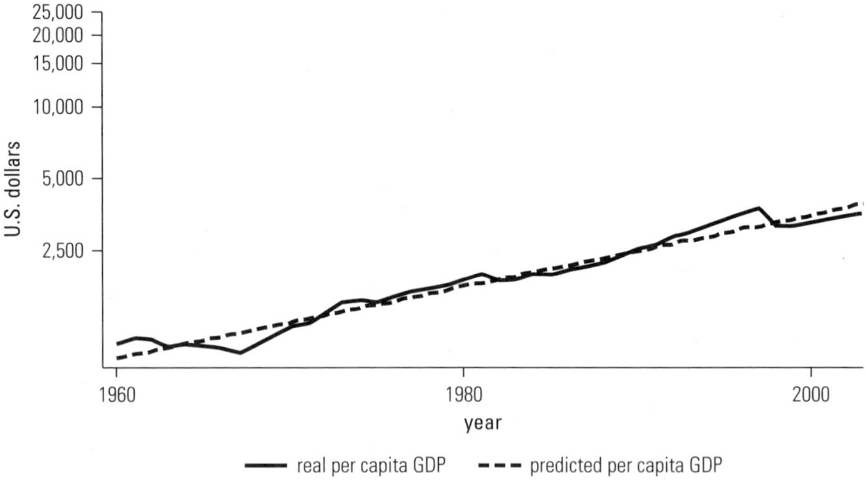

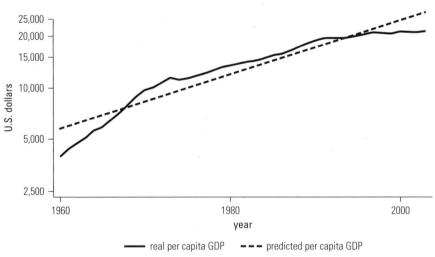

Source: Maddison 2006.

Figure 8.5 Per Capita GDP Growth of Japan, 1960–2003

Source: Maddison 2006.

Long-term trend growth rates for selected East Asian economies are also fairly stable, except for Japan (see figures 8.4–8.10). And as in the case of the European countries and the United States, all these economies grew faster between 1960 and 2000 relative to the fitted trend growth rate (see table 8.1). The exceptions were

Figure 8.6 Per Capita GDP Growth of the Republic of Korea, 1960–2003

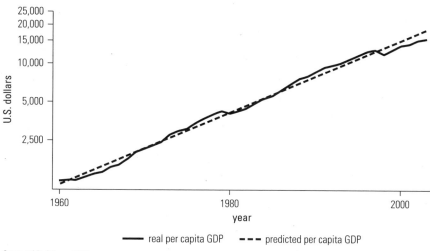

Source: Maddison 2006.

Figure 8.7 Per Capita GDP Growth of Malaysia, 1960–2003

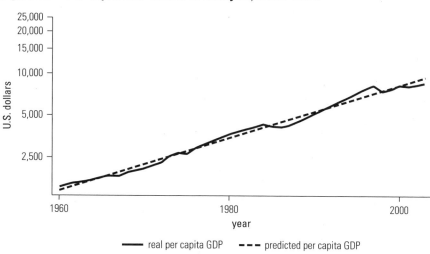

Source: Maddison 2006.

Indonesia and the Philippines. All the countries dropped below the trend line after 2000, and the brief acceleration in 2007 quickly lost steam after mid-2008.

As noted, R&D expenditure rose markedly from the 1960s onward as a share of GDP in Germany, the United Kingdom, and the United States, and it stabilized

Figure 8.8 Per Capita GDP Growth of the Philippines, 1960–2003

Source: Maddison 2006.

Figure 8.9 Per Capita GDP Growth of Taiwan, China, 1960–2003

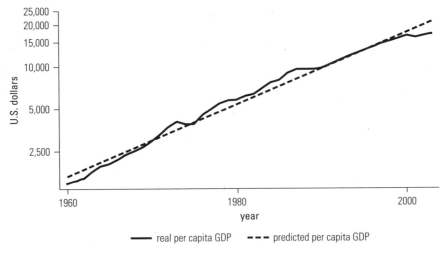

Source: Maddison 2006.

after 1985. Since the 1980s, R&D spending by Japan; Korea; and Taiwan, China, has also increased significantly (see table 8.2).

The stocks of human capital (which complement R&D spending), measured by the average years of schooling, have risen steadily in these economies since 1960. In Korea, for example, average years of schooling increased dramatically, from 4.25 years in 1960 to 10.80 years in 2000 (see figure 8.11).

Figure 8.10 Per Capita GDP Growth of Thailand, 1960–2003

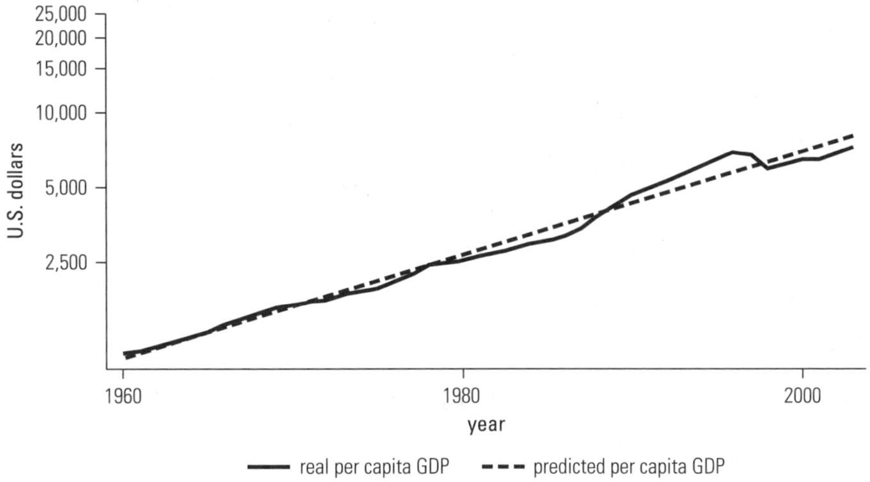

Source: Maddison 2006.

Table 8.2 R&D Expenditure, 1985–2006

Economy	R&D expenditure (as a % of GDP)				
	1985	1990	1995	2000	2006
Germany	2.6	2.6	2.2	2.5	2.5
United States	2.8	2.6	2.5	2.7	2.6
OECD total	2.2	2.3	2.1	2.2	2.3
China	—	—	0.6	0.9	1.4
Japan	2.8	3.0	2.9	3.0	3.4
Korea, Rep. of	—	—	2.4	2.4	3.2
Taiwan, China	—	—	1.7	2.0	2.6

Source: OECD 2007. Main Science and Technology Indicators, vol. 2007 release 02 (Data available at http://oberon.sourceoecd.org/vl=286308/cl=21/nw=1/rpsv/~3954/v207n1/s1/p1).
Note: — = not available.

An increase in enrollment in and completion of postsecondary education is responsible for the higher average years of schooling. In general, completion rates are about 50 percent in these economies. The exact timing of the rapid increase in tertiary education varies by economies, but most countries in the sample experienced a large increase in enrollment and completion of postsecondary education (see figures 8.12–8.16). By comparison, such an expansion in postsecondary education has yet to occur in Malaysia and the other Southeast Asian countries, except for the Philippines (see figures 8.17–8.20). The percentage of the population with some postsecondary education in the Philippines is well above that in

(*Text continues on page 199.*)

Figure 8.11 Average Years of Schooling, 1960–2000

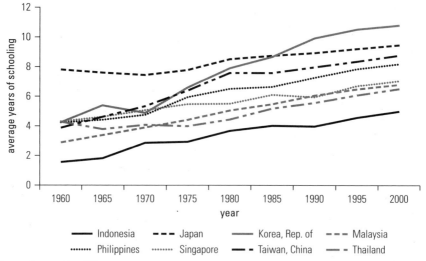

Source: Barro and Lee 2000.

Figure 8.12 Per Capita GDP Growth and Labor Force with Tertiary Education: Germany, 1860–2000

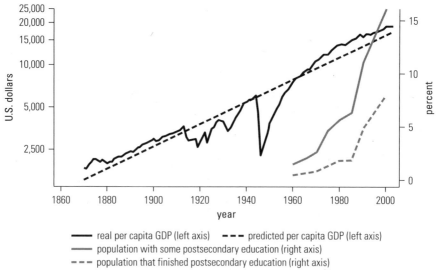

Sources: Barro and Lee 2000; Maddison 2006.

Figure 8.13 Per Capita GDP Growth and Labor Force with Tertiary Education: Japan, 1960–2000

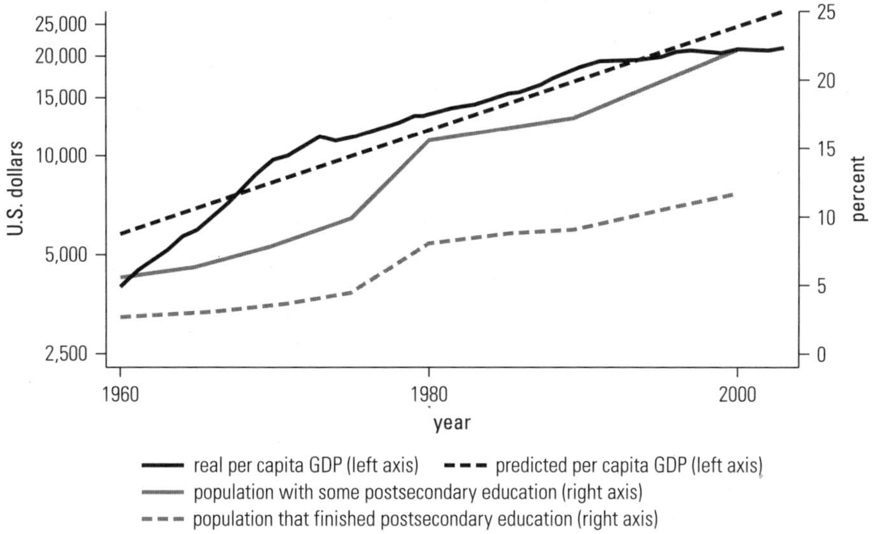

25,000
20,000
15,000
10,000
5,000
2,500
U.S. dollars
25
20
15
10
5
0
percent
1960
1980
2000
year
real per capita GDP (left axis)
predicted per capita GDP (left axis)
population with some postsecondary education (right axis)
population that finished postsecondary education (right axis)

Sources: Barro and Lee 2000; Maddison 2006.

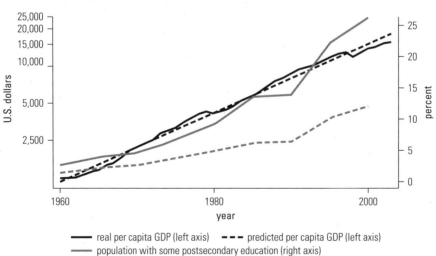

Figure 8.14 Per Capita GDP Growth and Labor Force with Tertiary Education: Republic of Korea, 1960–2000

25,000
20,000
15,000
10,000
5,000
2,500
U.S. dollars
25
20
15
10
5
0
percent
1960
1980
2000
year
real per capita GDP (left axis)
predicted per capita GDP (left axis)
population with some postsecondary education (right axis)
population that finished postsecondary education (right axis)

Sources: Barro and Lee 2000; Maddison 2006.

Figure 8.15 Per Capita GDP Growth and Labor Force with Tertiary Education: Taiwan, China: 1960–2000

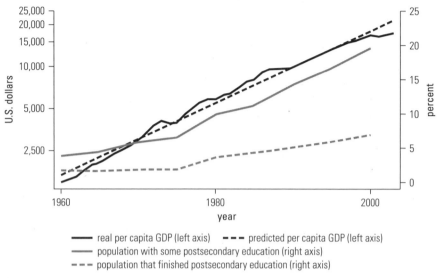

Sources: Barro and Lee 2000; Maddison 2006.

Figure 8.16 Per Capita GDP Growth and Labor Force with Tertiary Education: United States, 1860–2000

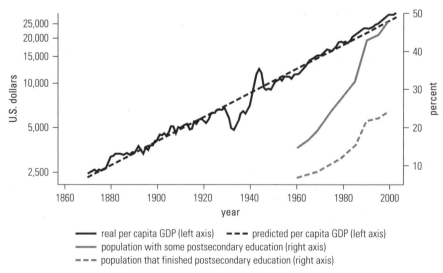

Sources: Barro and Lee 2000; Maddison 2006.

Can the Tigers Grow Fast and Furious Again? **197**

Figure 8.17 Per Capita GDP Growth and Labor Force with Tertiary Education: Indonesia, 1960–2000

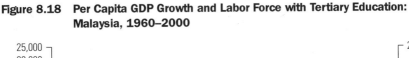

— real per capita GDP (left axis)   - - - predicted per capita GDP (left axis)
— population with some postsecondary education (right axis)
- - - population that finished postsecondary education (right axis)

Sources: Barro and Lee 2000; Maddison 2006.

Figure 8.18 Per Capita GDP Growth and Labor Force with Tertiary Education: Malaysia, 1960–2000

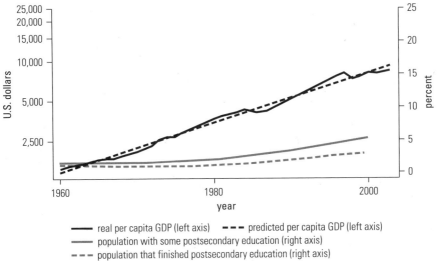

— real per capita GDP (left axis)   - - - predicted per capita GDP (left axis)
— population with some postsecondary education (right axis)
- - - population that finished postsecondary education (right axis)

Sources: Barro and Lee 2000; Maddison 2006.

Figure 8.19 Per Capita GDP Growth and Labor Force with Tertiary Education: Philippines, 1960–2000

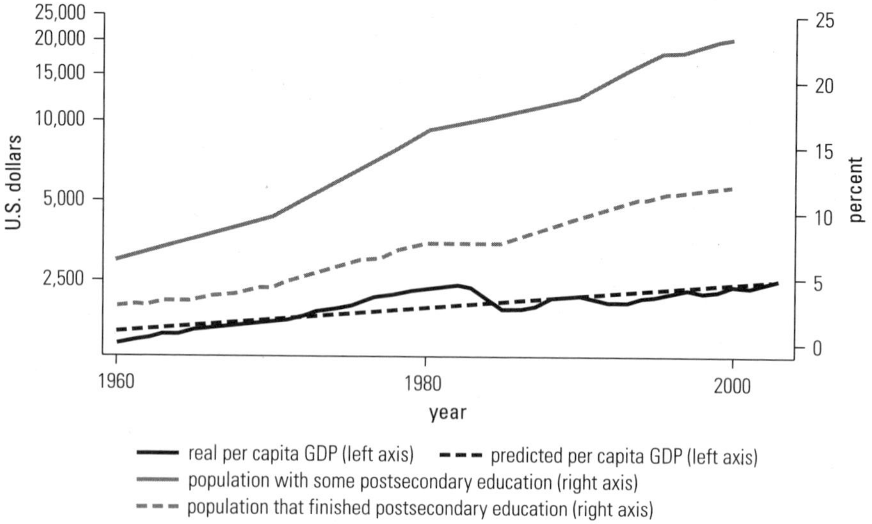

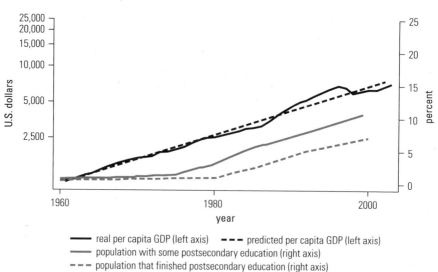

Sources: Barro and Lee 2000; Maddison 2006.

Figure 8.20 Per Capita GDP Growth and Labor Force with Tertiary Education: Thailand, 1960–2000

Sources: Barro and Lee 2000; Maddison 2006.

(*continued from page 193*)

the other three Southeast Asian countries; however, its growth rate has stubbornly resisted the pull that human capital has exerted in other countries in conjunction, no doubt, with other factors absent in the Philippines.

How Neighboring Economies Can Affect Malaysia

This analysis does not account for the effects of external factors on the growth of Southeast Asian countries—for example, changes in the level of investment and the industrial productivity of their principal trading partners. Determining how developments in other countries affect growth performance requires more complex modeling, and for this purpose, the applied general equilibrium model developed by the Global Trade Analysis Project (GTAP) is a handy tool.[2] It covers 57 industries in 66 regions and captures the interrelations among entities through two sets of equations: one set of accounting relationships ensures a balance between the receipt and expenditures of all agents, and a second set specifies the optimizing behavior of the agents involved (Brockmeier 2001). Two recent papers (Ianchovichina, Ivanic, and Martin 2009; Tongzon 2007) simulate the effects on Malaysian export values and employment of changing sectoral productivities in other countries, changes in investment levels in other countries, and changes in the quality of exports from other countries.

Productivity Change

From various model simulations, it appears that improvements in the productivity of China's export sector exert the most influence on Malaysia's export performance. Analysis by Tongzon (2007) indicates that a 100 percent increase in productivity in China would result in declines in exports of textiles (95 percent); wearing apparel (44 percent); electronic equipment (35 percent); chemical, rubber, and plastics (26 percent); oilseeds (21 percent); wood products (20 percent); and metal products (15 percent). Some business services, such as financial services (2.0 percent) and insurance (3.7 percent), would increase; others, such as other business services (0.7 percent) and recreation (1.2 percent), would decline. Overall, the effect of rising production in China would be a reduction in Malaysian exports of US$6.6 billion (about 5 percent of current exports). Effects stemming from other countries are broadly similar, but not as large and wide ranging as those from China. Productivity gains in Indonesia would mainly affect exports of oilseeds. Similarly, Ianchovichina, Ivanic, and Martin (2009) find that light manufactures,

---

[2] Although the findings are only suggestive, GTAP modeling can provide indicative orders of magnitude. In Malaysia's case, for example, a variant of the GTAP model estimated by Tongzon (2007) offers some clues as to how changes in Malaysia's trading partners can affect its sectoral performance.

machinery and equipment, apparel, and electronics would be affected most severely, with declines of 19 percent, 8 percent, 8 percent, and 7 percent, respectively, in Malaysian exports. The electronics and machinery subsectors would also see the largest price drops because of intensifying competition.

Employment in the textile, apparel, and electronics subsectors would be affected by increasing productivity in China and would decline by 78 percent, 31 percent, and 22 percent, respectively. These declines would be counterbalanced by increased employment in business services. Similarly, a productivity increase in Korea would negatively affect employment in the goods sector but would benefit services.

These results are based on changes in individual countries. Were all the partner countries to experience a 100 percent increase in productivity, the effect on Malaysia would inevitably be greatly amplified.

Investment

Were Malaysia's trading partners to increase investment by 100 percent, the effects on Malaysia would be much milder, compared with the case analyzed previously. Even if investment increased in China, the most affected subsector in Malaysia—textiles—would experience only an 8.2 percent decline in exports. The machinery and metal subsectors would suffer from falling exports if investment in Singapore or Thailand increased by 100 percent. An increase in investment in Singapore or Thailand would lead to a decline of 4.6 percent or 2.4 percent, respectively, in the output of the machinery subsector and of 3.9 percent or 2.1 percent, respectively, in the metals subsector.

If all countries were to raise their investments simultaneously, Malaysia's exports of oilseeds, textiles, and metal products would decline, whereas Malaysia would see an increase in the production of chemicals, electronic components, automotive parts, paper products, wearing apparel, and wood products. In general, the effect on employment would be muted because of the smaller magnitude of effects on exports.

Quality Improvement

When other countries raise the quality of their products (measured by a 100 percent increase in the unit price of a commodity), the outcome is favorable for Malaysia. As with changes in productivity, quality enhancement by China has the greatest effect on exports from Malaysia because Malaysian producers can then expand their share of the lower-quality product where they have an established advantage. In this scenario, Malaysia's textile exports would increase by 132 percent, followed by wearing apparel, with a 94 percent increase. The magnitudes of the positive effects from other countries would be smaller.

However, if all the countries (except Malaysia) were to improve product quality because of a shift in industrial technological thresholds at the same time, the consequences for Malaysia would be substantially negative, especially in the electronic equipment and machinery subsectors. If all countries (including Malaysia) were to see quality improvements, Malaysia's exports would decline substantially across the board (relative prices with respect to others do not change). This finding suggests that Malaysia cannot compete on the basis of low cost (because the quality improvements are measured by changes in unit prices).

Overall, the GTAP simulations indicate that even if world trade revives fully, Malaysia would have difficulty resuming export-led growth momentum on the basis of price competitiveness alone.[3] The other Southeast Asian countries confront a very similar predicament, though perhaps less acutely than Malaysia because their per capita GDPs are lower.[4] Furthermore, depending on the assumed changes, some Malaysian firms could shift their focus partially to the domestic market from export markets. Although doing so might increase employment, the gains would be short term until the small domestic market became saturated.

The growth performance of China—and to a lesser extent India—will benefit Malaysia's exports of agricultural and resource-based goods. But unless Malaysia can make its high-tech industry more competitive through investments in human and physical capital, its exports of electronics and machinery will be negatively affected, in terms of both the quantity exported and the prices for these goods (Ianchovichina, Ivanic, and Martin 2009).[5]

The GTAP models are useful because they embrace interaction and feedback effects among countries and offer suggestive numerical estimates. The model simulation available for Malaysia reinforces the message of the study: to sustain its exports, Malaysia will need to enhance its competitiveness by investing in the latest vintages of equipment, by taking measures to raise its productivity, and by building its innovation capacity.

---

[3] While the simulations presented here assume 100 percent changes, the qualitative results are not altered if 25 percent and 50 percent changes are used. The magnitude of the effect changes proportionally to the size of the shock.

[4] Some countries in the Greater Mekong region might be able to promote their trade through investment in transport infrastructure and in trade facilitation policies, which reduce transaction costs (Stone and Strutt 2009).

[5] In addition, improvements in logistics efficiency will benefit the primary and resource-based exports (Ianchovichina, Ivanic, and Martin 2009).

9

What Can the Tigers Do?

Policy makers in Malaysia and elsewhere in Southeast Asia have been conscious of the need to accelerate technology development from at least the mid-1990s, when export growth of some standardized manufactures slowed and alarm bells began ringing throughout the region. From then onward, Malaysia and Thailand, in particular, have begun introducing a wide spectrum of incentives (a) to encourage research and development (R&D) and technology acquisition by firms; (b) to promote the acquisition and upgrading of labor skills; (c) to facilitate the entry of new firms that could be the conduits for innovation; (d) to encourage multinational corporations (MNCs) to transfer technology and increase domestic sourcing of inputs; (e) to expand research in public institutions; (f) to increase the supply of science, technology, engineering, and math skills from universities; (g) to raise the quantum of research done in universities; and (h) to multiply links between universities and the business sector. Malaysia also began attempting to attract Malaysian knowledge workers who had gone overseas to return to their home country.

Some of the policy measures, fiscal or otherwise, are listed in tables 9.1 and 9.2. Suffice it to say, they are broad ranging and comparable in scope to those of other leading economies in East Asia (see appendix E for policies adopted by the Chinese government and appendix F for Thailand). Undoubtedly, there is room to coordinate the fiscal incentives among agencies and to make it easier for firms to access incentives, to improve the business climate by diminishing transaction costs, and to raise the standards of teaching and research in tertiary institutions. The current incentives in each of these areas have been described in recent reports by the World Bank (2005b, 2006a, 2007c), and ways to strengthen them have been proposed in some detail. It almost goes without saying that the development of a knowledge economy with significant innovation capabilities is inseparable from an increase in

Table 9.1 Malaysia's Incentive Policies

Policy	Agency or program	Objective	Year	Outcome
Business climate and technological development	Malaysian Business Council, Malaysian Technology Development Corporation, and Malaysia Industry Government High Technology Council	To facilitate information exchange, business incubation, and investment management	1993	The policy led to various initiatives to improve the quantity and quality of R&D.
Human Resource Development Act (1992)	Human Resource Development Council, which was formed to control the grants for training under the Human Resource Development Fund	To provide incentives for employers to upgrade the capacities and capabilities of their employees	1993	By 1997, the fund dispersed RM 99 million to train 533,227 people. At the end of 2002, the total number of employers registered was 8,172, of which 5,797 were in manufacturing and 2,375 in services. Between 1993 and 2002, electronics components, electrical machinery, and apparatus appliances and supplies producers were the largest group of industries. Between 1995 and 2002, hotel, freight, and computer were the top 3 industries in the services sector. In 2006, 608,962 employees were trained, involving financial assistance amounting to RM 2.89 billion. A total of 130,725 employees were trained in the areas of productivity and quality.
Higher education: student support mechanisms	National Higher Education Fund Corporation	To offer subsidized loans to help students meet the high tuition fees charged by the newly established private higher education institutions	1997	Between 1997 and 2005, about RM 15.1 billion was committed to almost 800,000 students.
Tertiary education: quality assurance Stimulation of skills-upgrading, technology acquisition, and R&D	National Accreditation Board and Quality Assurance Division, Technology Acquisition Fund and Intensification of Research in Priority Areas	To help Malaysian firms seek strategic technology from foreign sources and to fund R&D in Malaysian R&D institutions	1997	Public research institutes, such as the Malaysian Institute of Microelectronic Systems (MIMOS) and the Standards and Industrial Research Institute of Malaysia (SIRIM), promote basic and early-stage R&D in budding technology sectors and supply development assistance to local firms.

Policy	Agency	Objective	Year	Outcome
Technology transfer	Vendor Development Program	To offer incentives to more technologically advanced firms, usually foreign MNCs, to provide local companies with guaranteed contracts and a free interchange of engineers and product specifications	1993	
Domestic investment	Small and Medium Industry Development Corporation	To coordinate all incentives and assistance for the technological development of local firms	1995	The policy provides a 50% training grant to small and medium-size enterprises that send their employees to courses. Twenty-two training providers were appointed to undertake the Skills Upgrading Program for small and medium-size enterprises.
Upgrade technology: Third Industrial Master Plan	Multimedia Development Corporation and Multimedia Super Corridor	To improve technological competitiveness by shifting focus toward higher-value-added activities, including efforts to promote and support industrial clusters	1996–97	A venture capital fund and the Multimedia University were established. More than 1,100 firms were located in the Multimedia Super Corridor.
Second Science and Technology Policy (STEP2)	Industry R&D Grant Scheme, Commercialization of R&D Fund Multimedia Super Corridor R&D Grant Scheme, and Demonstrator Application Grant Scheme	To encourage firms to adapt and create new technologies To promote the commercialization of R&D results To promote the development of R&D clusters among Multimedia Super Corridor–status companies with at least 30% Malaysian equity To encourage diffusion of information and communication technology into the community	2003	Thirteen initiatives in human resource development were established. They aim to increase national R&D spending to at least 1.5% of gross domestic product by 2010.
Liberalization of manufacturing foreign direct investment (Tham 2004)	Ministry of International Trade and Industry	To achieve faster recovery from the crisis	1998–2003	The program allowed 100% foreign equity ownership in the manufacturing sector for new and expansion or diversification projects that applied by December 31, 2003 (except projects in metal stamping, metal fabrication, wire harnessing, printing, paper and plastic packing, plastic injection molded components, and steel service centers). It also removed export conditions.

Source: Data provided by the Economic Planning Unit of Malaysia.

Table 9.2 List of Major Investment Incentives Available for the Manufacturing Sector

Tax incentive	Description
Pioneer status	Income of high-tech companies and companies in an approved industrial linkages scheme is exempt for 5 years. Thereafter, a 30% corporate income tax rate applies. In Sabah, Sarawak, Labuan, and the designated Eastern Corridor of Peninsular Malaysia, the rate is 15%.
Investment tax allowance (ITA)	An allowance is granted of 60% (80% for Sabah, Sarawak, Labuan, and the designated Eastern Corridor of Peninsular Malaysia) of qualifying capital expenditure incurred during the first 5 years. The allowance can be used to offset 70% (85% for Sabah, Sarawak, Labuan, and the designated Eastern Corridor of Peninsular Malaysia and 100% for high-technology companies) of statutory income in the year of assessment. Any unused allowance can be carried forward to the following year until the amount has been used up. For companies specializing in R&D activities, an allowance is granted of 100% for R&D, contract R&D, and technical or vocational training companies (50% for in-house R&D companies) with respect to qualifying capital expenditures incurred during the first 10 years. The allowance can be used to offset 70% of statutory income in the year of assessment. Any unused allowance can be carried forward to the following year until the amount has been used up.
Reinvestment allowance (RA)	An allowance is granted of 60% of capital expenditure incurred by companies. The allowance can be used to offset 70% (100% for Sabah, Sarawak, Labuan, and the designated Eastern Corridor of Peninsular Malaysia and companies that can improve significantly in productivity) of statutory income in the year of assessment. The allowance is given for a period of 5 years beginning from the year in which the first reinvestment is made. After expiry of RA, companies producing promoted products or engaging in promoted activities are eligible for an accelerated capital allowance on capital expenditure by which 40% of the initial rate and 20% of the annual rate will enable capital write-off within 3 years.
Incentives for industrial adjustment	Incentives are given to the manufacturing sector for reorganization, reconstruction, or amalgamation within the same sector; enhancement of industrial self-sufficiency; improved industrial technology, increased productivity, and enhancement of efficient use of personnel and resources.
Incentives to strengthen the Industrial Linkages Scheme, incentives for large companies, and incentives for vendors	Tax deductions are provided for expenditure incurred for employee training and product development. Pioneer status or ITA status is granted for 5 years, with an exemption of 100% exemption of statutory income.
Incentives for export	Incentives grant a double deduction for promotion of exports, a double deduction on freight charges, and a double deduction on export credit insurance premiums. A tax exemption applies to the value of increased exports, and an industrial building allowance and export credit-refinancing scheme are available.
Incentives for promoting Malaysian brand names	A double deduction is granted for expenditure on local advertisements. Provisional fees are paid to companies promoting Malaysian brand names.

Table 9.2 (continued)

Tax incentive	Description
Training incentive on preemployment training: double deductions for expenses incurred on approved training under the Human Resources Development Fund	A single deduction is allowed for training expenses incurred before the commencement of business. Double deductions are allowed on expenses incurred for employee training at approved training institutions.
Infrastructure allowance	Companies that are engaged in the manufacturing or commercial sector in East Malaysia and the Eastern Corridor that incur expenses on qualifying capital infrastructure are given an infrastructure allowance of 100%. The allowance can be used to offset 85% of statutory income in the year of assessment. Any unused allowance can be carried forward to the following year until the amount has been used up.
Incentives for research and development contracts	Companies are eligible for pioneer status with a full income tax exemption at the statutory level for 5 years or for an ITA of 100% on qualifying capital expenditure within 10 years. The ITA can be used to offset 70% of the statutory income in the year of assessment. Companies are eligible to apply for an ITA of 100% on qualifying capital expenditure incurred within 10 years. The ITA can be used to offset 70% of the statutory income in the year of assessment. Companies are eligible to apply for an ITA of 50% on qualifying capital expenditure within 10 years.

Source: Adapted from Tham 2004.

the quality and volume of technical skills.[1] The World Bank's (2005b) report on the investment climate in 2002 and on firms' competitiveness in Malaysia identified a shortage of skilled labor[2] and regulatory burdens, such as the difficulty in hiring foreign workers, uncertainties arising from complex bureaucratic procedures related to obtaining licenses, and inefficiencies in business support services, as impediments to doing business. Preliminary findings from the second survey of the investment climate in 2008 indicated that skill shortages were less acute and, more surprisingly, that production was becoming less skill intensive.[3] Nevertheless, in the interests of

---

[1] See "Capturing Talent" (2007) and Guarini, Molini, and Rabellotti (2006) on parallel efforts by the Republic of Korea to raise labor productivity.

[2] The survey revealed that 50 percent and 80 percent of firms had vacancies for professional and skilled production workers, respectively, and it typically took them more than six weeks to fill such vacancies. In addition, a 25 percent wage premium is paid to college graduates relative to high school graduates, suggesting that the supply of skills is far short of the demand (Udomsaph and Zeufack 2006). The survey for 2008 pointed to a significant easing of labor market pressures.

[3] This development, which can be ascribed to the increased capital intensity of production and the adaptation of production technologies to long-standing skill shortages, may constrain the productivity growth of Malaysian industry. Some see the evolution of technology being influenced more by factors and skills in industrializing countries as more production shifts to these economies. Kim and Shafi'i (2009) found that deteriorating technical efficiency, which is highly dependent on the skills and quality of labor, is a contributing factor to slowing total factor productivity growth in Malaysia.

building a knowledge economy that enlarges the scope for innovation, there is no alternative to increasing the supply of graduates with tertiary education. This effort must go hand in hand with better instruction that raises students' proficiency in English and their communication, team-working, and information technology (IT) skills. Scaling up the skills development institutes of the caliber of the Penang Skills Development Centre and simplifying the process of applications and approval from human resource development funds would strengthen the national innovation system by promoting collaboration between firms and institutions. A favorable environment of incentives and a clearer definition and enforcement of intellectual property rights would be valuable complements.

To raise the quality of higher education, Malaysia must increase the efficiency of the public resources invested in universities and the accountability for this spending (World Bank 2007c). Concurrent with funding research competitively and selectively, providing better career prospects and higher salaries for potential students would also attract greater private resources from the citizens (and firms). For instance, innovation within the university system could be promoted by establishing professionally managed technology commercialization offices housed in selected universities, by developing technology broker programs, and by involving universities in regional development efforts. Eliminating tax holidays such as the pioneer status, reducing corporate income taxes, and maintaining investment-related allowances would strengthen fiscal incentives in favor of R&D and innovation (World Bank 2006a). However, the fiscal implications of such changes would need to be evaluated, and the effects of policies on FDI, for example, would have to be tracked to determine whether outcomes conform to expectations.

From this analysis of development in Malaysia and its neighbors, as well as from the experience of small, innovative, and successful economies—Finland; Israel; Switzerland; and Taiwan, China—a number of policy options emerge.[4]

Small economies in the middle- and higher-income categories need to specialize in a limited number of tradable products and services, which will deepen their comparative advantages in these areas, while always remaining on the lookout for opportunities to diversify into profitable new niches, most of which are likely to be offshoots of their current mix of products and services. Steadily augmenting comparative advantage in chosen areas and then leveraging the acquired capabilities to diversify, as the opportunities arise, into neighboring subfields would appear to be the optimal strategy (Hayes and Pisano 1994) and one favored by entrepreneurs, who typically start firms in industries with which they are familiar.[5]

---

[4] See Routti (2003) for a summary of Finland's knowledge economy development strategy; Kim (2000) on technological borrowing and knowledge economy development in Korea; and Jan and Chen (2006) on how interaction between government research institutes and firms has promoted technology development in Taiwan, China.

[5] In analyzing the lackluster growth of Peru since the 1980s, Hausmann and Klinger (2008) find that the inability of Peru to diversify into more promising exports was the primary cause of economic stagnation.

Industrial development, which is keyed to the existing comparative advantage plus a strategy for evolving this comparative advantage, is more likely to deliver favorable long-term results. Malaysia's apparent comparative advantage resides in products and services associated with electronics, agricultural biotechnology, food processing, and—possibly down the road—specialized financial services. Advantages are currently less apparent in the metallurgical, chemical, automotive, textile, engineering, and pulp and paper industries. The intensification of basic and applied research and of technology absorption efforts by firms, universities, and research institutes in a few areas is more likely to lead to technological upgrading and innovation.

Malaysia's growth strategy should focus on enhancing the potential of mid-technology, resource-based industries—in particular, wood products, food processing, and palm oil and its byproducts. Indonesia and, to a lesser extent, Thailand could also take this route. The innovation revealed comparative advantage (IRCA) can be thought of as a potential leading indicator for future exports. That is, strong innovation capabilities in one area should, over time, translate into competitive exports. Comparing the list of products in the IRCA for Malaysia and the list of products in later revealed comparative advantage (RCA) tabulations reveals that, thus far, only the innovation comparative advantage in amplifiers has led to an RCA in exports (see tables 9.3 and 9.4). In other fields, Malaysia's IRCA is not echoed in the RCA at a later date. Such disjointedness is also seen in Thailand (see tables 9.5 and 9.6). This finding suggests that the innovation capabilities remain disassociated from commercialization and exports, reinforcing concerns about the lack of focus in R&D and in the effort to commercialize innovations.

Given Malaysia's strong production base in electronics and electrical engineering and continuing investment by MNCs in these two subsectors, the longer-term growth prospects of this industry in Malaysia remain promising if Malaysian firms can move up the value chain through localization, design and product innovation, and productivity enhancement by improving process technologies. Increased R&D in both the resource-based and electronics industries plus measures to facilitate entry by local entrepreneurs will be the key to future growth in per capita gross domestic product. A transfer of human and financial resources to these subsectors could be supported by measures that scale back the incentives for industries where Malaysia's comparative advantage remains uncertain after more than two decades (for example, in auto assembly and auto parts). A shakeout and restructuring of this sector would seem to be desirable, and the huge overcapacity in the auto industry arising from the global recession that commenced in 2008 might make this unavoidable.

Gains in productivity and technological innovation can also be promoted by a competition law backed by policies that increase the pressure on firms to consolidate and lower the barriers to the entry of new firms. Larger firms, on balance, are a more prolific source of incremental innovation, and they find it easier to build productive capacity and to market globally, whereas new entrants in certain industries can be a source of more radical product innovation.

Table 9.3 Malaysia: Highest IRCA in 1995 and RCA in 2000

Class title	IRCA 1995	Product name	RCA 2000
Horizontally supported planar surfaces	105.8	Logs, meranti (light or dark red), bakau	57.36
Bottles and jars	93.5	Lumber, meranti red, meranti bakau, white lauan, and so forth	44.31
Geometrical instruments	45.4	Thorium ores and concentrates	40.85
Conveyors, power-driven	44.4	Palm oil or fractions simply refined	40.53
Illumination	26.9	Turntables with automatic record-changing mechanisms	39.74
Telecommunications	24.9	Palm kernel and babassu oil, fractions, simply refined	39.04
Stock material or miscellaneous articles	7.4	Cinematographic cameras for film less than 16 millimeters wide	38.55
		Cinematographic cameras for film greater than 16 millimeters wide	34.71
		Profile projectors, not elsewhere specified	32.89
		Palm nut or kernel oil cake and other solid residues	32.83
		Harpsichords, keyboard stringed instruments, not elsewhere specified	31.14
		Rubber gloves, other than surgical	29.52
		Calculating machines, nonelectric	27.22
		Glycerol (glycerin), crude and glycerol waters and lye	26.03
		Rubber surgical gloves	24.44
		Unglazed ceramic mosaic tiles and so forth less than 7 centimeters wide	24.07
		Vulcanized rubber thread and cord	23.06
		Dodecan-1-ol, hexadecan-1-ol, and octadecan-1-ol	22.63
		Plywood, 1 or 2 outer ply nonconifer, not elsewhere specified (ply less than 6 millimeters)	22.05
		Veneer or ply sheet, tropical woods, less than 6 millimeters thick	22.03

Sources: U.S. Patent and Trademark Office (USPTO); United Nations Commodity Trade Statistics Database (UN Comtrade).

There are opportunities in services industries, such as tourism and logistics. Hospitals in Penang have sought to promote medical tourism, mainly from Indonesia and the Middle East, with a degree of success. Many countries in Asia are vying for the medical tourist, and this factor must temper expectations. Medical services certainly have high value, but unless local links are developed, the employment multiplier and technological spillovers are small. Competition in logistics is also likely to be fierce (depending on future trend rates of growth in long-distance trade), and in the future, the returns could be smaller, could be subject to greater uncertainty, and might take longer to accrue.

Table 9.4 Malaysia: Highest IRCA in 2000 and RCA in 2005

Class title	IRCA 2000	Product name	RCA 2005
Conveyors, fluid current	306.1	Logs, meranti (light or dark red), bakau	63.5
Static molds	107.1	Turntables with automatic record-changing mechanisms	55.9
Glass manufacturing	21.4	Lumber, meranti red, meranti Bakau, white Lauan, and so forth	52.5
Expanded, threaded, driven, headed, tool-deformed, or locked-threaded fastener	18.8	Cameras for recording microfilm and so forth	51.6
Solid material comminution or disintegration	17.8	Cassette players, nonrecording	41.6
Etching a substrate, processes	17.4	Palm kernel and babassu oil, fractions, simply refined	40.9
Stoves and furnaces	16.8	Thorium ores and concentrates	40.7
Chemistry of hydrocarbon compounds	16.2	Palm oil or fractions, simply refined	36.5
Oscillators	14.0	Cigars, cheroots, cigarettes, with tobacco substitute	33.5
Geometrical instruments	11.9	Rubber gloves, other than surgical	32.1
Electrical transmission or interconnection systems	11.5	Radio-broadcast receivers, not elsewhere specified	30.0
Electricity, circuit makers and breakers	11.0	Instant print film, rolls	28.3
Beds	10.6	Dodecan-1-ol, hexadecan-1-ol, and octadecan-1-ol	26.1
Dentistry	10.4	Vulcanized rubber thread and cord	25.8
Amplifiers	8.9	Stearic acid	25.6
Conveyors, power-driven	8.9	Record players without built-in loudspeaker, not elsewhere specified	25.6
Aeronautics and astronautics	8.6	Ultramarine and preparations based thereon	25.1
Data processing, generic control systems or specific applications	8.1	Electrical control and distribution boards, greater than 1 kilovolt	24.2
Communications, directive radio wave systems and devices (such as radar, radio navigation)	7.6	Assembled battery watch movement, not elsewhere specified	23.8
Surgery, light, thermal, and electrical application	7.3	Industrial fatty alcohols	23.5

Sources: USPTO; UN Comtrade.

Enhancing comparative advantage in selected areas calls for concentrating public resources allocated for research in universities and research labs on a narrower front. Malaysia's R&D activities are scattered, and the results have been below par. This is also the case in other Southeast Asian countries. Some institutional changes could help strengthen and focus the research effort and increase the likelihood of findings being commercialized by small and medium-size

Table 9.5 Thailand: Highest IRCA in 1995 and RCA in 2000

Class title	IRCA 1995	Product name	RCA 2000
Combustion	43.3	Natural rubber in smoked sheets	77.24
Bleaching and dyeing; fluid treatment and chemical modification of textiles and fibers	30.4	Manioc (cassava) starch	76.85
Beds	25.0	Watch movements, complete and assembled, not elsewhere specified	74.27
Pipe joints or couplings	20.4	Natural rubber in other forms	73.66
Registers (for example, cash registers, calculators, devices for counting movements of devices)	13.0	Flour or meal of sago, starchy roots, or tubers	62.03
Dispensing (apparatus and process)	12.9	Rice flour	52.5
Refrigeration	10.5	Diamonds, industrial, unworked or simply sawn, cleave	48.53
Surgery (includes class 600)	4.5	Natural rubber latex, including prevulcanized	48.45
		Shrimp and prawns, prepared or preserved	47.7
		Rice, broken	47.68
		Cotton yarn, less than 85% single uncombed, less than 125 decitex, not retail	46.83
		Washing machines, not elsewhere specified, capacity less than 10 kilograms, built-in dryer	43.39
		Record players with loudspeakers, not elsewhere specified	40.33
		Pineapples, otherwise prepared or preserved	34.75
		Kapok	33.33
		Tuna, skipjack, bonito, prepared or preserved, not minced	32.32
		Rice, semimilled or wholly milled	31.06
		Ceramic troughs (agriculture), ceramic pots, and so forth	30.22
		Crispbread	30.05
		Thorium ores and concentrates	28.35

Sources: USPTO; UN Comtrade.

enterprises (SMEs).[6] Malaysia might well consider a body such as Finland's national technology agency, Tekes, which reports to the Ministry of Trade and Industry and is responsible for planning, coordinating, and financing natural technology programs. It also promotes technology transfer and runs technology clinics for SMEs (Sabel and Saxenian 2008).

Norway's TEFT (Technology Transfer from Research Institutes to SMEs) program is a useful model for encouraging research by SMEs, providing them with the supporting infrastructure for their research, and assisting them step by step in accessing technology suitable for their needs and customizing it for commercial

---

[6] However, the government must be alert against providing research grants to underachieving small firms that then use those resources to survive and continue producing (Lerner 2002). Such assistance affects the profitability of more efficient firms in the subsector, aside from wasting resources.

Table 9.6 Thailand: Highest IRCA in 2000 and RCA in 2005

Class title	IRCA 2000	Product name	RCA 2005
Fire extinguishers	45.2	Natural rubber in other forms	74.9
Sewing	26.8	Manioc (cassava) starch	70.1
Coating implements with material supply	15.6	Natural rubber latex, including prevulcanized	64.75
Ships	12.2	Cotton yarn, less than 85% single uncombed, less than 125 decitex, not retail	64.33
Metal fusion bonding	9.5	Natural rubber in smoked sheets	59.31
Music	9.1	Rice flour	52.26
Cleaning compositions for solid surfaces, auxiliary compositions therefor, or processes of preparing the compositions	6.6	Washing machines, not elsewhere specified, capacity less than 10 kilograms, built-in dryer	50.45
Multicellular living organisms and unmodified parts thereof and related processes	6.3	Rice, broken	44.91
Image analysis	5.5	Pineapples, otherwise prepared or preserved	40.49
Supports (for holding articles, for example)	5.1	Tuna, skipjack, bonito, prepared or preserved, not minced	36.54
Communications, electrical	2.3	Diamonds, industrial, unworked or simply sawn, cleave	33.69
Active solid-state devices (such as transistors, solid-state diodes)	1.3	Cotton sewing thread, less than 85% cotton, not retail	32.11
Semiconductor device manufacturing, process	0.8	Radio receivers, external power, not sound reproducers	31.07
Drug, bio-affecting and body-treating compositions (includes class 514)	0.5	Coin or disk-operated record players	26.85
		Vulcanized rubber thread and cord	25.52
		Rice, semimilled or wholly milled	25.19
		Woven cotton, not elsewhere specified, less than 85% manmade fiber, less than 200 grams, unbleached	24.64
		Pineapple juice, not fermented or spirited	24.5
		Refrigerators, household type, including nonelectric	24.11
		Engines, spark-ignition reciprocating, less than 50 cubic centimeters	23.87

Sources: USPTO; UN Comtrade.

purposes. The Fraunhofer institutes in Germany can be models for an important element of such a supporting infrastructure. The 58 institutes that emerged from the Fraunhofer model of 1973 employ a large number of scientists and engineers and engage mainly in contract research for private firms or public bodies. The funding mechanism and the relative autonomy enjoyed by the institutes have induced an entrepreneurial approach to applied research that has helped to ignite interest among SMEs, stimulate R&D effort by industry, and underpin the research endeavors of firms in Germany and elsewhere. Six Fraunhofer institutes operate in the

United States, and three operate in Asia. By establishing a couple of satellite research institutes overseas to serve as "technology sentinels" with strong links to domestic institutions, Malaysia could open windows for smaller firms and for budding entrepreneurs.

Side by side, MNCs could be induced by more stringently enforced conditions to transfer technology in priority areas to Malaysian firms.[7] Such localization was more practicable in the 1980s. Now the balance of power in the Malaysian context resides with the MNCs, which can more credibly threaten to relocate their operations elsewhere. The likelihood of a global redistribution of manufacturing industries has increased since 2008, and Southeast Asian countries may have a harder time attracting or even holding on to foreign direct investment.

Large Malaysian firms and MNCs that produce in areas where Malaysia enjoys a comparative advantage and views the strengthening of this advantage as a longer-term objective will be the standard-bearers of a more innovative economy. Much of the technology transfer associated with catch-up will be done by firms. Most of the applied research, technology, development, and commercialization will also be done by firms (with smaller firms complementing innovation by large firms in areas such as software and biotechnology). In the absence of such leadership by technology-oriented and internationally successful firms, Malaysia's ambitions with regard to growth and innovation will not be realized. Large firms will necessarily have to work in close cooperation with local research, training, financial, and other institutions. Absent their initiative and the strategic ambition of Malaysian firms to become major players in the global economy, the growth strategy will soon run out of steam.

It is the large firms that must take the lead in three other areas that are crucial to the development of technology capabilities in Malaysia. First, these firms need to create systems for improving the skills of their employees and providing attractive career paths for their ablest managerial and technical personnel, so as to minimize turnover, which is high but not critical, as it is in India, for example. The successful Japanese, Korean, and Western firms have depended not merely on their capacity to recruit qualified workers but, much more important, on longer-term, in-house personnel development; rotational and retention strategies, which enhanced the firm's human and research capital; and the capacity to engage in incremental innovation through the joint efforts of the research, production, and marketing departments. How this was and is done by firms such as Canon, General Electric, Google, HSBC, Procter & Gamble, Samsung, and Toyota is no secret. These firms prioritize innovation, recruit talented individuals, continuously add to their skills, stimulate them with challenges, and then handsomely reward their

---

[7]Almeida and Fernandes (2008) find that MNCs are less likely to transfer their latest technologies to wholly owned subsidiaries. Although most foreign companies in Malaysia are wholly owned, it is difficult to say whether their parent companies are slow to transfer innovation.

achievements (Ready and Conger 2007). Teaching and training institutions, of course, have their roles in raising the quality of the workforce, but in today's world, if firms are to find the mix and quality of skills they need to participate effectively in the global marketplace, the responsibility rests more with the firms themselves to orchestrate and, to a significant extent, finance the training internally and through the services of other providers.[8] In a highly competitive environment, firms cannot afford to wait for the government to provide the skills on tap; a cooperative effort is needed, with the business sector taking the lead in both defining and helping to implement tertiary-level curricula development and education strategies, which Universiti Sains Malaysia and Multimedia University are increasingly taking to heart. A supply-side approach to high-level human resource development largely directed by the government is wasteful and clearly not yielding the sought-after results and is resulting in labor market mismatches. As long as the business sector adopts a passive role and does not attach the requisite importance to human talent, the shortage of specialized skills and the problem of quality will not go away. It is only when employers actively engage with the authorities and with the providers of tertiary-level skills and shoulder more of the decision making, as well as the financial responsibilities, that the right mix and quality of human capital will be forthcoming.[9] This problem is regionwide, with other Southeast Asian countries also being pinched by shortages of skilled and technical workers with well-grounded technical and problem-solving skills. More flexible procedures to accommodate the immigration of knowledge workers may be needed to help absorb this technology (Rasiah 2008; World Bank 2008c).

Second, firms need to show more initiative in seeking out appropriate technology from all possible sources worldwide and buying or licensing it. In the past 10 years, numerous IT-based companies have emerged to play the role of intermediaries between the inventors of technology and potential users. These intermediaries search for solutions and serve as matchmakers between firms and suppliers of technology. They can complement a firm's own efforts and greatly expand the menu of options. Malaysian firms are doing too little of this scouting and acquiring of technologies, as

---

[8] A double-deduction incentive for training (DDIT) was introduced in Malaysia in the late 1980s. However, it had limited success. Between 1987 and 1993, 591 in-house training programs with 3,253 trainees were approved. The Human Resource Development Fund (HRDF) replaced the DDIT for firms with 50 employees or more. The HRDF is a levy scheme. It allows firms that have contributed 1 percent of total payroll for six months to claim a portion of training expenditures. In 2000, RM 833 million was collected and RM 488 million (59 percent) was disbursed (Fleming and Søborg 2002).

[9] Even in the United States, the supply response to an increase in wage premium has been small. The wage premium of U.S. college graduates increased by more than 20 percent, but the supply response was only about 1 percent after the educational background of parents and other household characteristics were controlled for (Altonji, Bharadwaj, and Lange 2008).

is apparent from the limited outlay for licensing and royalties. Here, the government research institutes can be more proactive in creating their own search mechanisms, as Eli Lilly did with InnoCentive; can use the existing providers of search and transfer services; or can set up their own mechanisms for provoking technical suggestions from respondents across the world on improving production processes and creating better products (see Nambisan and Sawhney 2007).

Upgrading a company's human resources and stepping up efforts to borrow ideas and technologies from others are inseparable from increasing in-house research. Assimilation and adaptation of technology acquired elsewhere, reverse engineering of products, late-stage modification of technology for commercialization, and customization of products to suit the requirements of specific niche markets all depend on in-house capabilities of a certain level. Identifying promising technologies that others have devised is difficult unless the firm knows approximately what it is looking for and how it can be mated with its own research, product development, or process technologies. Moreover, as the firm moves closer to the technology frontier, borrowing becomes less passive. It begins to require greater capacity to adapt and innovate until, eventually, firms such as Samsung and Canon must devise most of the key innovations themselves or in partnership with other firms that are also at the technological frontier. Many of Malaysia's leading firms are nowhere close to the technological frontiers in their respective fields, and acquiring innovation capability is a long, drawn-out process, especially when it comes to building research skills; a cadre of experienced project managers; links with or investment in other firms with desirable intellectual property; and a firm-level system for stimulating, producing, and marketing innovations. Innovative firms are not built overnight. The examples of Hyundai, Samsung, Sony, Taiwan Semiconductor Manufacturing Company, and now China International Marine Containers, Infosys, Lenovo, Wipro, and others show how lengthy and resource intensive the process can be. No Malaysian firm—or for that matter no firm in Indonesia, the Philippines, or Thailand—has as yet embarked on the path to achieving competitiveness that is based on innovation. Until a few firms make this commitment, governments are relatively powerless, and more incentives for developing human resources will largely be futile.

Malaysia's IT infrastructure needs strengthening to maximize productivity gains for manufacturing and services activities and to stimulate innovation. Upgrading the infrastructure will help increase labor productivity. It can also enhance the agglomeration economies and technological spillovers in cities. Although the costs of such upgrading are considerable, the knowledge economy of urban areas is increasingly dependent on the quality of the information and communication technology infrastructure. Achieving parity with the frontrunners in the East Asia region is a necessary objective for cities such as Jakarta, Indonesia; Kuala Lumpur, Malaysia; and Bangkok, Thailand.

The final suggestion is arguably the archstone that anchors the others. Malaysia's knowledge economy will likely never emerge from four or five midsize cities that are weakly connected to one another and to the leading global knowledge centers. Currently, even the largest Malaysian urban area, which is the Kuala Lumpur–Klang Valley urban region,[10] lacks the scale, heterogeneity, and innovation culture needed to give rise to significant agglomeration economies, to create a dense and diverse market for skilled workers, and to generate technological spillovers. Policies that support a more decentralized urban development will compound the problems of smallness. By spreading resources more thinly, the potential for assembling the critical masses of human capital and other factors in enabling urban settings is reduced further.[11] Just as there is a need for focus in R&D, there is an equal need for focused urban development that underpins the knowledge economy. Small, isolated urban-industrial islands have proved to be ineffective foundations for an innovative knowledge economy in Malaysia and in other countries (for example, Australia). An alternative strategy might be warranted—one that seeks to grow networked industrial clusters in urban centers that have the scale to give rise to agglomeration economies. Whether Malaysia concentrates its industrial and knowledge economy resources in one or two strategically located urban areas remains to be determined. However, a nation of 27 million people with a small stock of skilled workers will find it difficult to achieve the growth through a deepening of technological capabilities if it attempts to cultivate multiple urban centers in the hope of developing four or five local innovation systems that, over time, can compete with those in the most dynamic Asian and European economies.

The upgrading of technology and the building of an innovation system must be coordinated with macroeconomic policies and with policies favoring the Malay majority introduced in 1970 by Tun Abdul Razak, the father of Najib Razak.[12] Thus, measures that encourage productive investment and sustain commensurate rates of domestic savings are an integral part of a growth strategy, as are measures that maintain the revenue effort that backstops incentives for R&D spending by public entities and universities. Over the longer term, the experience of Korea; Singapore; and Taiwan, China also shows that exchange rate policy has a key role. In each of these cases, a variety of intersecting policies have contributed collectively to the upgrading of technology and exports, and each of these economies struggles ceaselessly to sustain a mix of policies conducive to the dynamics of industry (see also Mahadevan 2007a).

---

[10] See Bunnell, Barter, and Morshidi (2002) on the ongoing efforts to make Kuala Lumpur into a global city.

[11] The Northern Corridor Economic Region (NCER) is the second of four corridors to be launched in the Ninth Plan (2006–10). The plan for the NCER spans four national plans (up to 2025), yet it has been allocated only RM 5 billion of government funding so far, and reception by the private sector is cool. In contrast, the Iskandar Development Region (the southern corridor) has attracted a large amount of private investment ("Malaysia: Northern Project" 2007).

[12] King (2007) offers a similar recipe for the Philippines.

The partial dismantling of the pro-Malay policies announced by Prime Minister Najib Razak in mid-2009 signals an awareness of the threat suspended over the Malaysian Tiger. By more than halving the equity stake allotted to ethnic Malays in most companies and by permitting foreign investors to hold a larger ownership share in finance, education, and health care organizations, the Malaysian government has demonstrated a readiness to grasp the nettle and to begin easing what some have perceived as a serious constraint on entrepreneurship, industrial change, and innovation in Malaysia (Burton 2009b).

Appendix A
Revealed Comparative Advantage of East Asian Economies Other than Malaysia

Table A.1 Top 10 Commodities with Highest Revealed Comparative Advantage in China, 2007

Product	Product name	RCA	PRODY	Technology classification
580123	Woven weft pile cotton fabric, not elsewhere specified, width less than 30 centimeters	9.77	8,974	LT1
610323	Men's and boys' ensembles of synthetic fibers, knit	9.37	6,020	LT1
500200	Raw silk (not thrown)	9.36	2,861	PP
670420	Wigs, false beards, eyebrows, and so forth, of human hair	9.24	7,710	LT2
610322	Men's and boys' ensembles of cotton, knit	9.21	6,020	LT1
580134	Woven warp pile fabric, manmade fibers, epingle (uncut)	9.17	12,873	MT2
580133	Woven weft pile fabric, manmade fibers, not elsewhere specified	9.00	12,873	MT2
610423	Women's and girls' ensembles of synthetic fibers, knit	8.98	6,020	LT1
360410	Fireworks	8.87	8,501	MT2
610422	Women's and girls' ensembles of cotton, knit	8.86	6,020	LT1

Source: United Nations Commodity Trade Statistics Database (UN Comtrade).
Note: HT1 = electronic and electrical products; HT2 = other high-technology products; LT1 = textiles, garments, and footwear; LT2 = other low-technology products; MT1 = automotive products; MT2 = process industry; MT3 = engineering products; PP = primary products; RB1 = agro-based products; RB2 = other resource-based products; RCA = revealed comparative advantage. Technology classification is based on Lall (2000). For each product, a weighted average of the gross domestic product per capita of countries exporting that product is calculated to assign a value (PRODY) that is a proxy for quality.

Table A.2 Top 10 Commodities with Highest Revealed Comparative Advantage in Indonesia, 2007

Product	Product name	RCA	PRODY	Technology classification
480710	Paper, laminated with bitumen tar or asphalt, uncoated	109.51	11,754	RB1
140210	Kapok	109.51	5,977	PP
310320	Basic slag, in packs, greater than 10 kilograms	109.50	6,023	MT2
010120	Asses, mules, and hinnies, live	109.31	16,901	PP
480521	Paper, multi-ply, each layer bleached, uncoated, not elsewhere specified	109.29	16,642	RB1
330121	Essential oils of geranium	109.18	3,705	RB2
480230	Paper, carbonizing base, uncoated	108.73	16,642	RB1
640330	Footwear, wood base, leather uppers, no inner sole	108.54	7,765	LT1
340410	Artificial and prepared waxes, of modified lignite	106.46	17,539	MT2
920910	Metronomes, tuning forks, and pitch pipes	106.20	13,528	LT2

Source: UN Comtrade.
Note: See table A.1.

Table A.3 Top 10 Commodities with Highest Revealed Comparative Advantage in Japan, 2007

Product	Product name	RCA	PRODY	Technology classification
392073	Sheet or film, not cellular or reinforced cellulose acetate	16.38	20,868	MT2
284610	Cerium compounds	14.29	8,320	HT2
700490	Drawn glass in sheets, clear	12.41	7,149	RB2
911011	Complete movements of watches, unassembled or partly assembled	12.31	11,985	MT3
845921	Numerically controlled metal-working drill machines	12.18	14,455	MT3
290612	Cyclohexanol, methylcyclohexanol, dimethylcyclohexanol	11.81	10,815	MT2
840721	Outboard motors, spark ignition	11.29	15,147	MT3
540339	Yarn of artificial filament, single, not elsewhere specified, not retail	11.28	9,493	LT1
871140	Motorcycles, spark ignition engine of 500–800 cubic centimeters	10.79	9,110	MT1
370199	Photographic plates, film in the flat, not elsewhere specified	10.77	24,756	MT2

Source: UN Comtrade.
Note: See table A.1.

Table A.4 Top 10 Commodities with Highest Revealed Comparative Advantage in the Republic of Korea, 2007

Product	Product name	RCA	PRODY	Technology classification
283719	Cyanides and cyanide oxides of metals, except sodium	18.88	9,141	RB2
890120	Tankers	15.51	14,263	MT3
845020	Household or laundry-type washing machines, capacity greater than 10 kilograms	14.32	15,565	MT3
290244	Mixed xylene isomers	13.64	15,052	RB2
480990	Paper, copying or transfer, width greater than 36 centimeters	13.30	22,143	RB1
890590	Floating docks, special function vessels, not elsewhere specified	12.88	10,562	MT3
890520	Floating, submersible drilling or production platform	11.87	10,562	MT3
901380	Optical devices, appliances and instruments, not elsewhere specified	11.26	20,604	HT2
845129	Drying machines, not elsewhere specified	11.16	15,565	MT3
401310	Inner tubes of rubber for motor vehicles	10.95	10,500	RB1

Source: UN Comtrade.
Note: See table A.1.

Table A.5 Top 10 Commodities with Highest Revealed Comparative Advantage in the Philippines, 2007

Product	Product name	RCA	PRODY	Technology classification
900911	Electrostatic photocopiers, direct process	341.16	17,269	HT1
481430	Wallpaper, covered one side with plaiting material	282.05	11,754	RB1
930610	Cartridges for rivet and similar tools, humane killers, and so forth	278.46	13,133	MT3
530521	Abaca fiber, raw	265.36	3,366	RB1
283090	Sulfides of metals, not elsewhere specified, polysulfides of metals	236.44	10,364	RB2
230650	Coconut or copra oil-cake and other solid residues	183.86	5,718	PP
151311	Coconut (copra) oil crude	179.18	3,322	RB1
741022	Foil, copper alloy, backed, thickness less than 0.15 millimeters	175.39	9,728	PP
80110	Coconuts, fresh or dried	163.72	1,727	PP
902740	Exposure meters	150.73	18,162	HT2

Source: UN Comtrade.
Note: See table A.1.

Table A.6 Top 10 Commodities with Highest Revealed Comparative Advantage in Singapore, 2007

Product	Product name	RCA	PRODY	Technology classification
292250	Amino-alcohol-phenols and so forth, with oxygen function	30.53	18,337	RB2
911019	Rough movements of watches	23.16	11,985	MT3
293722	Halogenated derivatives of adrenal cortical hormones, bulk	19.95	25,207	HT2
854190	Parts of semiconductor devices and similar devices	18.73	11,030	HT1
854219	Monolithic integrated circuits, except digital	18.69	16,474	HT1
910899	Watch movements, complete and assembled, not elsewhere specified	18.24	17,433	MT3
854290	Parts of electronic integrated circuits and so forth	16.08	11,030	HT1
840710	Aircraft engines, spark ignition	14.43	6,179	MT3
852320	Unrecorded magnetic disks	13.80	22,263	LT2
840910	Parts for spark ignition aircraft engines	12.32	6,179	MT3

Source: UN Comtrade.
Note: See table A.1.

Table A.7 Top 10 Commodities with Highest Revealed Comparative Advantage in Taiwan, China, 2007

Product	Product name	RCA	PRODY	Technology classification
291523	Cobalt acetates	50.80	16,255	MT2
392072	Sheet or film, not cellular or reinforced vulcanized rubber	50.72	20,868	MT2
920920	Mechanisms for musical boxes	50.36	13,528	LT2
291534	Isobutyl acetate	47.94	16,255	MT2
920993	Parts of and accessories for musical instruments	46.26	13,528	LT2
800510	Tin foil (thickness less than 0.2 millimeter)	45.50	7,639	PP
482319	Gummed or adhesive paper, in strips or rolls	45.18	9,630	LT2
741600	Springs, copper	44.92	15,810	LT2
293610	Provitamins, unmixed	43.49	18,879	HT2
950360	Puzzles	42.88	14,329	LT2

Source: UN Comtrade.
Note: See table A.1.

Table A.8 Top 10 Commodities with Highest Revealed Comparative Advantage in Thailand, 2007

Product	Product name	RCA	PRODY	Technology classification
110814	Manioc (cassava) starch	72.49	8,693	RB2
071410	Manioc (cassava), fresh or dried	63.84	4,789	PP
511000	Yarn of coarse animal hair or of horsehair	61.82	10,260	LT1
520615	Cotton yarn, less than 85% single uncombed, less than 125 decitex, not retail	61.55	4,262	LT1
400121	Natural rubber in smoked sheets	58.94	1,169	PP
400110	Natural rubber latex, including prevulcanized	57.69	1,169	PP
400280	Mixtures of natural and synthetic rubbers	55.15	1,169	PP
400129	Natural rubber in other forms	52.71	1,169	PP
100640	Rice, broken	42.94	4,455	PP
200820	Pineapples, otherwise prepared or preserved	40.76	9,337	RB1

Source: UN Comtrade.
Note: See table A.1.

Appendix B
Product Space Analysis for Southeast Asian Economies

Table B.1 Top 20 Upscale Commodities with Highest Density in China, 2000–04

Product name	Product	Density	Technology classification	PRODY − EXPY
Other sound recorders and reproducers	7638	0.537294	MT3	4,765.33
Watches, watch movements, and cases	8851	0.536328	MT3	6,868.30
Television receivers, monochrome	7612	0.524965	HT1	5,388.19
Children's toys, indoor games, and so forth	8942	0.514865	LT2	3,764.92
Fabrics, woven, of continuous synthetic textile materials	6531	0.502063	MT2	5,276.61
Umbrellas, parasols, walking sticks	8994	0.498902	LT2	1,848.52
Tableware and other articles of other kinds commonly used for domestic or toilet purposes	6665	0.488781	LT2	960.89
Knitted or crocheted fabric, not elastic or rubberized	6551	0.487574	LT1	1,880.06
Optical instruments and apparatus	8710	0.483351	HT2	10,039.76
Knitted or crocheted fabrics of fibers	6552	0.482366	LT1	207.44
Fabrics, woven, containing 85% discontinuous synthetic fibers	6532	0.479233	MT2	2,378.30
Peripheral units, including control and adapting units	7525	0.478158	HT1	5,142.37
Baby carriages and parts	8941	0.476559	LT2	7,898.67
Other radio broadcast receivers	7628	0.472699	MT3	9,061.52
Microphones, loudspeakers, amplifiers	7642	0.472566	HT1	1,301.51
Pins and needles, fittings, base metal	6993	0.469479	LT2	1,548.98
Printed circuits and parts thereof	7722	0.468498	MT3	2,855.42

(continued)

Table B.1 (continued)

Product name	Product	Density	Technology classification	PRODY – EXPY
Fish, prepared or preserved, not elsewhere specified	0371	0.468357	RB1	210.26
Batteries and accumulators and parts	7781	0.465225	HT1	2,324.80
Spectacles and spectacle frames	8842	0.464037	MT3	11,993.38

Source: Authors' calculations.

Note: HT1 = electronic and electrical products; HT2 = other high-technology products; LT1 = textiles, garments, and footwear; LT2 = other low-technology products; MT1 = automotive products; MT2 = process industry; MT3 = engineering products; PP = primary products; RB1 = agro-based products; RB2 = other resource-based products. Technology classification is based on Lall (2000). For each product, a weighted average of the gross domestic product per capita of countries exporting that product is calculated to assign a value (PRODY) that is a proxy for quality. The final column shows the difference between PRODY and EXPY. A positive number means "upgrading" in a sense of exporting more sophisticated commodities relative to the overall export basket (see chapter 3).

Table B.2 Top 20 Upscale Commodities with Highest Density in Hong Kong, China, 2000–04

Product name	Product	Density	Technology classification	PRODY – EXPY
Watches, watch movements, and cases	8851	0.400925	MT3	5,054.59
Other sound recorders and reproducers	7638	0.393827	MT3	2,951.62
Television receivers, monochrome	7612	0.374227	HT1	3,574.48
Optical instruments and apparatus	8710	0.373830	HT2	8,226.05
Diodes, transistors and similar semiconductor devices	7763	0.362926	HT1	903.15
Printed circuits and parts thereof	7722	0.348849	MT3	1,041.71
Electronic microcircuits	7764	0.348082	HT1	4,095.70
Peripheral units, including control and adapting units	7525	0.340043	HT1	3,328.66
Knitted or crocheted fabric, not elastic or rubberized	6551	0.339144	LT1	66.35
Children's toys, indoor games, and so forth	8942	0.338593	LT2	1,951.21
Fabrics, woven, of continuous synthetic textile materials	6531	0.338147	MT2	3,462.90
Parts of and accessories suitable for office machines and automatic data processing machines and units	7599	0.337380	HT1	3,083.83
Other radio broadcast receivers	7628	0.336517	MT3	7,247.81
Spectacles and spectacle frames	8842	0.331844	MT3	10,179.67
Fabrics, woven, containing 85% discontinuous synthetic fibers	6532	0.329378	MT2	564.59
Umbrellas, parasols, walking sticks	8994	0.328767	LT2	34.81

Table B.2 *(continued)*

Product name	Product	Density	Technology classification	PRODY − EXPY
Synthetic, reconstructed precious, and semiprecious stones	6674	0.317434	RB2	1,707.32
Gramophones and record players, electric	7631	0.313948	MT3	3,619.13
Baby carriages and parts	8941	0.308653	LT2	6,084.96
Other electric valves and tubes	7762	0.308425	HT1	5,289.06

Source: Authors' calculations.

Table B.3 Top 20 Upscale Commodities with Highest Density in India, 2000–04

Product name	Product	Density	Technology classification	PRODY − EXPY
Building and monumental stone, worked	6613	0.419143	RB2	1,698.40
Travel goods, handbags, briefcases	8310	0.410645	LT1	1,636.54
Cotton fabrics, woven, bleached, mercerized	6522	0.402457	LT1	836.17
Ships, boats, and other vessels for breaking up	7933	0.400409	MT3	9,051.07
Articles commonly used for domestic purposes	6974	0.396076	LT2	1,123.31
Fabrics, woven, of continuous synthetic textile materials	6531	0.393158	MT2	7,702.95
Fabrics, woven, of silk, noil, or other waste silk	6541	0.389829	LT1	684.84
Tulle, lace, embroidery, ribbons, and other small wares	6560	0.382619	LT1	755.43
Fish, dried, salted, or in brine; smoked fish	0350	0.380520	RB1	980.55
Television receivers, monochrome	7612	0.374214	HT1	7,814.53
Fabrics, woven, containing 85% discontinuous synthetic fibers	6532	0.373227	MT2	4,804.64
Iron ore agglomerates (sinters, pellets)	2816	0.370889	RB2	2,256.20
Other tires, tire cases, inner tubes	6259	0.370720	RB1	2,361.54
Carpets, rugs, and so forth of other textile materials	6596	0.370388	LT1	815.53
Yarn containing 85%, by weight, synthetic fibers	6514	0.370117	LT1	2,160.20
Yarn of regenerated fibers, not for retail	6517	0.367897	LT1	1,355.25
Carpets, carpeting, rugs, mats, and matting	6594	0.367059	LT1	1,403.99
Household appliances, decorative articles	6978	0.366912	LT2	587.08
Fish, prepared or preserved, not elsewhere specified	0371	0.365884	RB1	2,636.60
Tires, pneumatic, new, of a kind used on motorcycles and bicycles	6254	0.365374	RB1	1,151.28

Source: Authors' calculations.

Table B.4 Top 20 Upscale Commodities with Highest Density in Indonesia, 2000–04

Product name	Product	Density	Technology classification	PRODY – EXPY
Manufactured goods, not elsewhere specified	8999	0.411184	LT2	413.05
Clothing accessories, knitted or crocheted	8472	0.386699	LT1	804.88
Fish, dried, salted, or in brine; smoked fish	0350	0.382502	RB1	1,821.23
Articles of apparel and clothing accessories	8482	0.373568	LT1	875.37
Fish, prepared or preserved, not elsewhere specified	0371	0.373415	RB1	3,477.29
Macaroni, spaghetti, and similar products	483	0.371355	RB1	425.01
Jackets, blazers of textile fabrics	8424	0.370867	LT1	496.54
Other sound recorders and reproducers	7638	0.359750	MT3	8,032.36
Suits, men's, of textile fabrics	8422	0.358835	LT1	613.93
Footwear	8510	0.356121	LT1	467.64
Mate	0742	0.349879	PP	532.95
Basketwork, wickerwork, and so forth, of plait	8997	0.348577	LT2	491.95
Registers, exercise books, notebooks	6423	0.344369	LT2	280.88
Radio broadcast receivers, portable	7622	0.344152	MT3	2,242.57
Cotton fabrics, woven, bleached, mercerized	6522	0.342755	LT1	1,676.85
Fruit otherwise prepared or preserved	0589	0.341507	RB1	2,039.54
Yarn of discontinuous synthetic fibers	6516	0.337361	LT1	13.40
Tableware and other articles of other kinds commonly used for domestic or toilet purposes	6665	0.336282	LT2	4,227.92
Travel goods, handbags, briefcases	8310	0.334792	LT1	2,477.22
Ornamental articles	8933	0.334392	LT2	2,001.72

Source: Authors' calculations.

Table B.5 Top 20 Upscale Commodities with Highest Density in the Republic of Korea, 2000–04

Product name	Product	Density	Technology classification	PRODY – EXPY
Other sound recorders and reproducers	7638	0.405849	MT3	2,833.39
Optical instruments and apparatus	8710	0.390575	HT2	8,107.82
Other electric valves and tubes	7762	0.385882	HT1	5,170.83
Television receivers, monochrome	7612	0.384971	HT1	3,456.25
Diodes, transistors, and similar semiconductor devices	7763	0.382488	HT1	784.92
Electronic microcircuits	7764	0.368050	HT1	3,977.47
Peripheral units, including control and adapting units	7525	0.364232	HT1	3,210.43
Parts of and accessories suitable for office machines and automatic data processing machines and units	7599	0.345741	HT1	2,965.60
Printed circuits and parts thereof	7722	0.343804	MT3	923.48
Fabrics, woven, of continuous synthetic textile materials	6531	0.342729	MT2	3,344.67
Other radio broadcast receivers	7628	0.339193	MT3	7,129.58
Polystyrene and its copolymers	5833	0.339044	MT2	5,094.06
Polycarboxylic acids and their anhydrides	5138	0.335609	MT2	329.88
Spectacles and spectacle frames	8842	0.327402	MT3	10,061.44
Parts of and accessories for musical instruments	8989	0.325650	MT3	1,031.96
Complete digital data processing machines	7522	0.324885	HT1	1,419.11
Fabrics, woven, containing 85% discontinuous synthetic fibers	6532	0.323884	MT2	446.36
Pile and chenille fabrics, woven, of manmade fibers	6539	0.322503	MT2	376.28
Weaving, knitting machines, for preparing yarn	7245	0.320506	MT3	3,423.70
Gramophones and record players, electric	7631	0.318422	MT3	3,500.90

Source: Authors' calculations.

Table B.6 Top 20 Upscale Commodities with Highest Density in Malaysia, 2000–04

Product name	Product	Density	Technology classification	PRODY – EXPY
Other sound recorders and reproducers	7638	0.271593	MT3	4,114.85
Electronic microcircuits	7764	0.269055	HT1	5,258.93
Peripheral units, including control and adapting units	7525	0.268164	HT1	4,491.89
Diodes, transistors, and similar semiconductor devices	7763	0.265796	HT1	2,066.38
Gramophones and record players, electric	7631	0.263565	MT3	4,782.36
Other radio broadcast receivers	7628	0.261722	MT3	8,411.04
Complete digital data processing machines	7522	0.257725	HT1	2,700.57
Off-line data processing equipment	7528	0.253919	HT1	5,742.17
Parts of and accessories suitable for office machines and automatic data processing machines and units	7599	0.252966	HT1	4,247.06
Printed circuits and parts thereof	7722	0.239698	MT3	2,204.94
Other electric valves and tubes	7762	0.236802	HT1	6,452.29
Radio broadcast receivers for motor vehicles	7621	0.230843	MT3	6,760.08
Polycarboxylic acids and their anhydrides	5138	0.227190	MT2	1,611.34
Polystyrene and its copolymers	5833	0.223921	MT2	6,375.52
Parts of apparatus of electrical machinery	7649	0.215535	HT1	4,492.79
Microphones, loudspeakers, amplifiers	7642	0.213478	HT1	651.03
Television picture tubes, cathode ray tubes	7761	0.210742	HT1	1,211.49
Invalid carriages, motorized or not motorized	7853	0.210714	MT1	2,489.54
Calculating machines, cash registers	7512	0.208308	HT1	4,724.50
Other electric machinery and equipment	7788	0.207180	HT1	3,416.94

Source: Authors' calculations.

Table B.7 **Top 20 Upscale Commodities with Highest Density in the Philippines, 2000–04**

Product name	Product	Density	Technology classification	PRODY – EXPY
Ornamental articles	8933	0.225681	LT2	866.70
Watches, watch movements, and cases	8851	0.197844	MT3	9,000.31
Fish, dried, salted, or in brine; smoked fish	0350	0.195180	RB1	686.22
Fish, prepared or preserved, not elsewhere specified	0371	0.192989	RB1	2,342.27
Piezoelectric crystals, mounted	7768	0.180778	HT1	2,597.22
Diodes, transistors, and similar semiconductor devices	7763	0.179294	HT1	4,848.87
Travel goods, handbags, briefcases	8310	0.176165	LT1	1,342.20
Electronic microcircuits	7764	0.175305	HT1	8,041.42
Fruit otherwise prepared or preserved	0589	0.172487	RB1	904.52
Statuettes and other ornaments	6666	0.172140	LT2	667.24
Peripheral units, including control and adapting units	7525	0.171251	HT1	7,274.38
Radio broadcast receivers, portable	7622	0.166526	MT3	1,107.55
Candles, matches, pyrophoric alloys	8993	0.166280	LT2	4.96
Parts of and accessories suitable for office machines and automatic data processing machines and units	7599	0.164980	HT1	7,029.55
Insulated, electric wire, cable, bars, strip	7731	0.164480	MT3	126.47
Other sound recorders and reproducers	7638	0.163965	MT3	6,897.34
Television receivers, monochrome	7612	0.163015	HT1	7,520.20
Printed circuits and parts thereof	7722	0.162871	MT3	4,987.43
Small wares and toilet articles, feather dusters	8998	0.156259	LT2	1,130.93
Resistors, fixed or variable, and parts	7723	0.155931	MT3	226.79

Source: Authors' calculations.

Table B.8 Top 20 Upscale Commodities with Highest Density in Singapore, 2000–04

Product name	Product	Density	Technology classification	PRODY – EXPY
Electronic microcircuits	7764	0.297282	HT1	2,775.61
Off-line data processing equipment	7528	0.287881	HT1	3,258.85
Parts of and accessories suitable for office machines and automatic data processing machines and units	7599	0.275051	HT1	1,763.74
Peripheral units, including control and adapting units	7525	0.270839	HT1	2,008.57
Complete digital data processing machines	7522	0.261141	HT1	217.25
Other radio broadcast receivers	7628	0.256255	MT3	5,927.72
Complete digital central processing units	7523	0.246098	HT1	5,720.41
Other electric valves and tubes	7762	0.244655	HT1	3,968.97
Epoxide resins	5826	0.240356	MT2	5,873.32
Chemical products and flashlight materials	8821	0.234901	MT2	12,079.51
Sulfonamides, sultones, and sultams	5157	0.233908	RB2	9,322.61
Mixtures of 2 or more odoriferous substances	5514	0.233411	RB2	159.37
Gramophones and record players, electric	7631	0.232442	MT3	2,299.04
Ethers, alcohol peroxides, ether peroxides	5161	0.232272	RB2	2,078.54
Calculating machines, cash registers	7512	0.229154	HT1	2,241.18
Parts of and accessories suitable for office machines and automatic data processing machines and units	7591	0.228881	HT1	6,634.51
Polystyrene and its copolymers	5833	0.228100	MT2	3,892.20
Orthopedic appliances, surgical belts	8996	0.227525	LT2	5,368.51
Gramophone records and similar sound recordings	8983	0.227294	LT2	8,565.07
Heterocyclic compounds; nucleic acid	5156	0.226730	RB2	11,877.89

Source: Authors' calculations.

Table B.9 Top 20 Upscale Commodities with Highest Density in Thailand, 2000–04

Product name	Product	Density	Technology classification	PRODY – EXPY
Piezoelectric crystals, mounted	7768	0.395952	HT1	856.58
Peripheral units, including control and adapting units	7525	0.385531	HT1	5,533.74
Other sound recorders and reproducers	7638	0.383195	MT3	5,156.70
Other radio broadcast receivers	7628	0.382146	MT3	9,452.89
Gramophones and record players, electric	7631	0.380857	MT3	5,824.21
Diodes, transistors, and similar semiconductor devices	7763	0.380165	HT1	3,108.23
Printed circuits and parts thereof	7722	0.378205	MT3	3,246.79
Watches, watch movements, and cases	8851	0.376449	MT3	7,259.67
Fish, prepared or preserved, not elsewhere specified	0371	0.373926	RB1	601.63
Electronic microcircuits	7764	0.371953	HT1	6,300.78
Fabrics, woven, of continuous synthetic textile materials	6531	0.367298	MT2	5,667.98
Radio broadcast receivers for motor vehicles	7621	0.366241	MT3	7,801.93
Tableware and other articles of other kinds commonly used for domestic or toilet purposes	6665	0.364880	LT2	1,352.26
Parts of and accessories suitable for office machines and automatic data processing machines and units	7599	0.362466	HT1	5,288.91
Invalid carriages, motorized or not motorized	7853	0.360321	MT1	3,531.39
Clocks, clock movements, and parts	8852	0.356258	MT3	1,811.48
Polystyrene and its copolymers	5833	0.354389	MT2	7,417.37
Fabrics, woven, containing 85% discontinuous synthetic fibers	6532	0.354364	MT2	2,769.67
Yarn containing 85%, by weight, synthetic fibers	6514	0.353579	LT1	125.23
Parts of apparatus of electrical machinery	7649	0.351069	HT1	5,534.64

Source: Authors' calculations.

Appendix C
Research and Development Spending by Private Firms in Malaysia

Table C.1 Breakdown of Private Research and Development Spending

Industry	Amount (RM)
Agriculture, hunting, and related service activities	51,118,422
Forestry, logging, and related service activities	892,870
Fishing, operation of fish hatcheries and fish farms, and service activities incidental to fishing	0
Mining of coal and lignite; extraction of peat	2,792,577
Extraction of crude oil and natural gas; service activities incidental to crude oil and natural gas extraction excluding surveying	150,467,659
Mining of uranium and thorium ores	0
Mining of metal ores	0
Other mining and quarrying	0
Manufacturing of food products and beverages	8,729,775
Manufacture of tobacco products	0
Manufacturing of textiles	591,178
Manufacturing of wearing apparel; dressing and dyeing of fur	1,142,935
Tanning and dressing of leather; manufacture of luggage, handbags, saddlery, harnesses, and footwear	178,562
Manufacture of wood and products of wood and cork, except furniture; manufacture of articles of straw and plaiting materials	611,994
Manufacture of paper and paper products	1,302,117
Publishing, printing, and reproduction of recorded media	0
Manufacture of coke, refined petroleum products, and nuclear fuel	5,963,724
Manufacture of chemicals and chemical products	34,294,038
Manufacture of rubber and plastic products	46,733,380
Manufacture of other nonmetallic mineral products	2,673,822
Manufacture of basic metals	2,119,897
Manufacture of fabricated metal products, except machinery and equipment	3,259,995
Manufacture of machinery and equipment, not elsewhere classified	13,963,658
Manufacture of office, accounting, and computing machinery	414,450,260

(continued)

Table C.1 *(continued)*

Industry	Amount (RM)
Manufacture of electrical machinery and apparatus, not elsewhere classified	117,765,109
Manufacture of radio, television, and communication equipment and apparatus	396,344,262
Manufacture of medical and precision optical instruments, watches, and clocks	7,255,023
Manufacture of motor vehicle, trailers, and semitrailers	646,281,449
Manufacture of other transport equipment	9,884,873
Manufacture of furniture; manufacturing not elsewhere classified	168,000
Recycling	641,290
Electricity, gas, steam, and hot water supply	0
Collection, purification, and distribution of water	0
Construction	1,830,237
Sale, maintenance, and repair of motor vehicles and motorcycles	0
Wholesale trade and commission trade, except of motor vehicles and motorcycles	1,155,745
Retail trade, except motor vehicles and motorcycles; repair of personal and household goods	0
Hotels and restaurants	0
Land transport; transport via pipelines	0
Water transport	0
Air transport	729,081
Supporting and auxiliary transport activities; activities of travel agencies	0
Post and telecommunications	52,173,738
Finance, except insurance and pension funding	0
Insurance and pension funding, except compulsory social security	0
Activities auxiliary to finance	0
Real estate activities	0
Renting of machinery and equipment without operator and renting of personal and household goods	173,857
Computer and related activities	18,017,399
Research and development	22,516,239
Other business activities	1,420,564
Public administration and defense; compulsory social security	2,223,200
Education	5,936,747
Health and social work	6,896,747
Sewage and refuse disposal, sanitation and similar activities	137,809
Activities of membership organizations, not elsewhere classified	0
Recreational, cultural, and sporting activities	0
Other service activities	713,099
Private households with employed persons	0
Extraterritorial organizations and bodies	0
Total	2,033,551,331

Source: Economic Planning Unit of Malaysia.

Appendix D
Index of Innovation Revealed Comparative Advantage

Table D.1 Innovation Revealed Comparative Advantage: Top 10 Technology Classes in China, 2007

Class	Class title	Index value
71	Chemistry: fertilizers	18.5
279	Chucks or sockets	18.0
104	Railways	9.9
450	Foundation garments	9.5
335	Electricity: magnetically operated switches, magnets, and electromagnets	8.0
377	Electrical pulse counters, pulse dividers, or shift registers: circuits and systems	7.5
38	Textiles: ironing or smoothing	7.5
720	Dynamic optical information storage or retrieval	7.0
361	Electricity: electrical systems and devices	6.8
439	Electrical connectors	6.7

Source: U.S. Patent and Trade Office (USPTO).

Table D.2 Innovation Revealed Comparative Advantage: Top 10 Technology Classes in India, 2007

Class	Class title	Index value
71	Chemistry: fertilizers	26.2
1	Classification undetermined	18.0
246	Railway switches and signals	11.1
504	Plant protecting and regulating compositions	10.7
377	Electrical pulse counters, pulse dividers, or shift registers: circuits and systems	10.7
293	Vehicle fenders	9.6
532	Organic compounds (includes classes 532 to 570)	7.1
326	Electronic digital logic circuitry	7.0
28	Textiles: manufacturing	6.9
708	Arithmetic processing and calculating (electrical computers)	6.2

Source: USPTO.

Table D.3 Innovation Revealed Comparative Advantage: Top 10 Technology Classes in Indonesia, 2007

Class	Class title	Index value
95	Gas separation: processes	192.99
84	Music	84.11
166	Wells (shafts or deep borings in the earth, for example, for oil and gas)	41.78
29	Metal working	25.23
128	Surgery (includes class 600)	19.83

Source: USPTO.

Table D.4 Innovation Revealed Comparative Advantage: Top 10 Technology Classes in the Philippines, 2007

Class	Class title	Index value
228	Metal fusion bonding	57.82
363	Electric power conversion systems	51.07
101	Printing	38.55
257	Active solid-state devices (for example, transistors, solid-state diodes)	9.16
324	Electricity: measuring and testing	8.67
606	Surgery (instruments)	7.95
235	Registers (for example, cash registers, calculators, devices for counting movements of devices)	7.68
701	Data processing: vehicles, navigation, and relative location	6.95
438	Semiconductor device manufacturing: process	5.26
439	Electrical connectors	4.06
250	Radiant energy	3.93
424	Drug, bio-affecting, and body-treating compositions (includes class 514)	1.56

Source: USPTO.

Table D.5 Innovation Revealed Comparative Advantage: Top 10 Technology Classes in Singapore, 2007

Class	Class title	Index value
228	Metal fusion bonding	20.6
105	Railway rolling stock	9.3
232	Deposit and collection receptacles	7.7
271	Sheet feeding or delivering	7.4
451	Abrading	7.0
330	Amplifiers	5.5
294	Handling: hand and hoist-line implements	5.4
257	Active solid-state devices (for example, transistors, solid-state diodes)	5.2
708	Arithmetic processing and calculating (electrical computers)	5.1
438	Semiconductor device manufacturing: process	5.1

Source: USPTO

Table D.6 Innovation Revealed Comparative Advantage: Top 10 Technology Classes in Thailand, 2007

Class	Class title	Index value
281	Books, strips, and leaves	1,787.3
363	Electric power conversion systems	46.4
379	Telephonic communications	15.9
399	Electrophotography	11.5
29	Metal working	11.5
361	Electricity: electrical systems and devices	7.6
250	Radiant energy	7.2
382	Image analysis	6.7
455	Telecommunications	4.1
438	Semiconductor device manufacturing: process	3.2
424	Drug, bio-affecting, and body-treating compositions (includes class 514)	2.8

Source: USPTO.

Appendix E
Financial Incentives for Research and Development, Technology Development, and Innovation in Chinese Firms

Table E.1 Financial Incentives for Innovation Offered in China

Type	Description
Fiscal incentives for research and development (R&D) and related activities	The Chinese government provides import tariff exemption: • To facilitate technological renovation and product upgrading in existing state-owned enterprises. In addition, targeted industries, such as those in the electronics sector, were exempted from tariffs and import-related value added tax on equipment during the 9th and 10th five-year periods. • To promote technical transfer and commercialization. Foreign individuals, firms, and R&D centers engaged in consulting activities and technical services related to technology transfer and technological development are exempt from corporate income tax.
Fiscal incentives in various technology development zones	Establishing economic zones, new and high-tech industrial zones (HTIZs), and economic and technological development zones is one of the key measures that the government has adopted in facilitating acquisition of new and advanced technologies, promoting technological innovation, promoting the commercialization of science and technology results, and enhancing China's industrial competitiveness. China began establishing special economic zones from the early 1980s and HTIZs from the 1990s. In 1991, China approved 21 national HTIZs, and by 2005, the total number countrywide had risen to 150, of which 53 are at the national level. These HTIZs have nursed 39,000 high-tech firms employing 4.5 million people. The total turnover of firms reached RMB 2.7 trillion in 2004, an increase of 31% over the previous year. The per capita profit was RMB 33,000, the per capita tax yield was RMB 29,000, and the per capita foreign earnings were RMB 157,320 (US$19,000). In the national HTIZs, a series of investor-friendly policies and measures have been introduced. These measures include tax reduction and exemption policies.
Fiscal incentives related to income tax	The Chinese government offers various tax holiday schemes to different types of firms: • Income tax: — Foreign-invested enterprises can enjoy income tax exemption in the first 2 years after making profits and an income tax reduction (by half) in the following 3 years. — Foreign-invested high-tech enterprises can enjoy income tax exemption in the first 2 years after making profits and an income tax reduction (by half) in the following 6 years. — Sino-foreign joint ventures can enjoy income tax exemption in the first 2 years after making profits. — Other firms are eligible for income tax exemption in the first 2 years from starting productive operation.

(continued)

Table E.1 *(continued)*

Type	Description
	— Domestic firms in HTIZs are eligible for preferential treatment but with limits in terms of types of business activities (income earned from technology transfer or activities related to technology transfer, such as technical consulting services and training). A ceiling is imposed on how much they can benefit from income tax exemption (less than RMB 300,000). — The corporate income tax rate is set at 15% in these zones, which is much lower than the rate for companies located outside the zones. Firms whose export share is above 70% of their annual production can enjoy a further income tax reduction (10%). • Turnover tax: — Foreign enterprises and foreign-invested enterprises are exempt from the business tax on technology transfer. • Tariff and import duties: — Tariff and import-stage value added tax exemptions have been granted to foreign-funded enterprises for their importation of equipment and technologies that are listed in the catalogue of encouragement. • Accelerated depreciation: — Since 1991, new and high-tech firms are granted accelerated depreciation for equipment and instruments (see China's State Council Document No. 12 of 1991).
Scholarships for students studying in science and engineering fields in China and abroad	The Chinese government has created the Overseas Study Fund to sponsor Chinese students and scholars pursuing their studies or training overseas. In 2004, the fund sponsored 3,630 people for advanced studies or research programs overseas. In line with China's development priorities, the fund identified 7 disciplines or academic fields as its sponsorship priorities for 2004: • Telecommunications and information technology • High and new technology in agricultural science • Life science and population health • Material science and new materials • Energy and environment • Engineering science • Applied social science and subjects related to World Trade Organization issues
Incentives to attract overseas Chinese to return to China	The *Chunhui* program has sponsored 8,000 Chinese scholars with PhDs obtained overseas to return to the country to carry out short-term work. The Yangtze River Fellowship program awarded 537 overseas Chinese scholars professional appointments in Chinese universities to build curricula, teach, and conduct joint academic research.
Fiscal incentives to attract the establishment of R&D centers by multinational corporations	The fiscal incentives offered under this program include the following: • Since 1997, exemption from import duties and import-related value added tax for imports of equipment, devices, and spare parts for R&D purposes. • Since November 2004, exemption from tariffs and import-related value added tax for acquiring imported new and advanced technologies. Foreign-funded R&D centers receive the same fiscal benefits as foreign-funded high-tech firms and enjoy the same fiscal preferential treatments. • Since 1999, exemption from corporate income tax on revenue earned through the delivery of consulting or other technical services related to technology transfer and revenue earned from technical development activities. • A corporate income tax reduction for those R&D centers whose expenditures on R&D increased more than 10% annually.

Source: Yusuf, Wang, and Nabeshima 2009.

Appendix F
Financial Incentives for Research and Development, Technology Development, and Innovation in Thai Firms

Table F.1 Support for Investments in the Development of Skills, Technology, and Innovation in Thailand

Scheme	Organization	Objective	Description	Supporting measures	Outcome
National Science and Technology Development Agency (NSTDA) Investment Center	NSTDA	To promote research and development (R&D) spending by the private sector in science and technology (S&T), with a focus on human resource development, capital funding, and S&T management	NSTDA coinvests in projects that support the national S&T policy, such as those that require advanced technology to create innovative products to reduce R&D risks of private firms. Projects must have the potential to be commercialized and have reasonable returns on investment. Projects must enhance value-added products that will reduce imports. Projects also must support the transfer of technology and preserve the environment.	NSTDA will fund less than 50% of the total investment. NSTDA will be part of the management team in accordance with its share of investment in the project. NSTDA will withdraw funding from the project if the project is determined to be ineffective or if its funding is no longer necessary.	
Investment Development Policy for Enhancing Technology and Innovation	Board of Investment	To stimulate and provide incentives for firms to improve their technology capabilities	The policy supports direct S&T investment in the following areas: • Manufacture of pharmaceutical and medical equipment • Manufacture of S&T equipment • Manufacture of aviation spare parts • Electronic designs • R&D • S&T testing services • Calibration • Human resource development	Firms are exempt from R&D machinery import duties. Firms receive a corporate income tax holiday for 1–8 years.	

Good Innovation–Zero-Interest Scheme	National Innovation Agency	To provide investment opportunities for the private sector to innovate by coabsorbing risks	The scheme provides soft loans for start-up firms to create prototype products or pilot projects.	The soft loans are issued by the National Innovation Agency and participating financial institutions. The maturity is less than 3 years. The firms are responsible for providing collateral.	In 2005, 22 projects were supported totaling B 23,650,000. The cumulative amount of support as of 2005 was B 1,172,500,000.
Technology Capitalization Scheme	National Innovation Agency	To support the private sector in applying knowledge to create new products or patents	The scheme provides grant support for distinguished innovation projects with a high degree of novelty.	The private sector must provide at least 25% of the total investment. Grants are for up to 75% of the total investment, not to exceed B 5,000,000 per project. The maturity is less than 3 years.	In 2005, 13 projects were supported totaling B 16,580,000. The cumulative amount of support as of 2005 was B 54,380,000.
Innovation Cluster Grant Scheme	National Innovation Agency	To promote private sector R&D clusters	The scheme provides grants for projects that promote clusters, such as manufacturing and regional clusters. Eligible projects range from pilot projects to commercialization programs.	The grant amount is for less than B 5,000,000 per project. The maturity is less than 3 years.	In 2005, 6 projects were supported totaling B 9,040,000. The cumulative amount of support as of 2005 was B 80,890,000.
Venture Capital Scheme	National Innovation Agency	To promote investments in industries with high potential	The National Innovation Agency and joint venture institutes will invest in the project an amount not to exceed 49% of the project's registered capital.	The National Innovation Agency must hold a smaller share than the joint venture institutes, and the total amount of the agency's investment may not exceed B 25,000,000.	Between 2004 and 2006, 6 projects were supported totaling B 39,500,000. The cumulative amount of support as of 2006 was B 325,000,000.

Source: World Bank 2008d.

Table F.2 Expanded Support Programs for Enhancing Technology in Industry in Thailand

Scheme	Organization	Objective	Description	Supporting measures	Outcome
Industrial Technology Assistance Program	National Science and Technology Development Agency	To set up a mechanism to form links between technology providers and users	The program provides technical experts who assist in undertaking research and development (R&D), provide consulting, and help solve problems at the factory location. Assistance includes • Matching local demand in technology with external suppliers • Providing in-country and overseas technology consultants to enhance levels of production and R&D • Organizing seminars in areas of technology that aim to enhance the capability of personnel in organizations • Searching for appropriate technology • Conducting quality assessments	Payments for experts to diagnose general technical problems are covered at full cost. Payments for hiring experts for technology development are covered at 50% of costs incurred, not to exceed B 500,000. Such support is limited to 2 projects per firm per year.	During 1992–2001, the program supported 630 projects from 562 firms.
Program Directing Companies for Technology Development	National Science and Technology Development Agency	To provide funds to the private sector for R&D conducted to improve products and production processes that are based on appropriate technology	The program provides soft loans for the following activities: • Conducting R&D and commercializing the findings • Improving technology or production processes and products • Setting up or upgrading research labs	The maximum loan is B 30,000,000, not to exceed 75% of the project's total cost. The interest rate is one-half of the general deposit rate in 1 year plus 2.25 percentage points. The payment period is 7 years, with no payments of principal in the first 2 years.	

Program	Agency	Objective	Description	Terms	Notes
Program Directing Companies for Technology Development to Improve Competitiveness	Department of Industrial Promotion	To develop industry throughout the value chain (production processes, quality assurance, R&D in products, and financial management and marketing) and to promote technology transfers from universities and research institutes to the private sector to enhance productivity at the firm level	Forty small and medium-size enterprises were selected to join with consultants in the following activities: • Developing and improving production processes • Improving standards and products so that they meet International Organization for Standardization 9000 standards • Enhancing capacity for planning, technology management, strategic planning, and marketing so that firms can compete in international markets	Providing financial support for 60% of consulting costs, not to exceed B 900,000.	In 2005, B 40,000,000 was allocated to the program. The program resulted in an increase in sales of participating firms of B 2,605,200,000.
Consultancy Fund	Department of Industrial Promotion	To provide consulting services to enhance productivity at the firm level	Program activities are as follows: • Hiring consultants to provide general supervision to the firms • Hiring consultants to provide technical diagnosis to the firms • Monitoring the firms	Funds are provided for 50% of consulting costs, not to exceed B 200,000 if the procurement procedure is through bids and not to exceed B 100,000 if the procurement procedure is direct selection.	In 2006, B 16,800,000 was allocated.
Knowledge Creation Fund	Office of the Higher Education Commission	To help support R&D investment by the private sector and government agencies	The program provides funds to projects related to knowledge creation and knowledge application.		The fund is currently being set up.

Source: World Bank 2008d.

Table F.3 Support under the Revised Policy on Intellectual Property in Thailand

Scheme	Organization	Objective	Description	Supporting measures	Outcome
Intellectual Property Services Program	National Science and Technology Development Agency	To encourage private sector R&D and to protect Thai property rights	The program provides services related to property rights to the private sector.	Services involve • Providing advice and consultations on the process of property rights application and property rights commercialization • Helping to coordinate searches for property rights information • Giving specialist advice and consultations on legal matters • Providing training and seminars on property rights issues	During 1999–2005, the program worked with the private sector on 46 property rights cases, 55 licensing cases, and 31 cases dealing with trademarks and other issues.
Cooperation on implementation in the areas of innovation and intellectual property	National Science and Technology Development Agency, National Innovation Agency, Department of Business Development, Export Promotion Department, and Intellectual Property Department	To coordinate cooperation among the 5 government agencies with respect to innovation creation and intellectual property protection and commercialization; to provide services in the areas of innovation, intellectual property, and Thai business promotion; to cooperate on setting up measures, procedures, and mechanisms for innovation creation and intellectual property protection and commercialization; and to sign a memorandum of understanding that allows the flow of information and the creation of openness among the agencies.	A memorandum of understanding has been signed by the 5 government agencies to demonstrate their commitment to working together.	A framework has been set up to promote cooperation in 6 areas: • Innovation creation and intellectual property • Property rights protection • Property rights commercialization • Property rights enforcement • Human resources development in innovation and intellectual property • Thai business promotion	

Source: World Bank 2008d.

References

Acharya, Ram C., and Wolfgang Keller. 2007. "Technology Transfer through Imports." NBER Working Paper 13086, National Bureau of Economic Research, Cambridge, MA.

Aghion, Philippe, and Peter Howitt. 2005. "Appropriate Growth Policy: A Unifying Framework." Harvard University, Cambridge, MA.

Ahlstrom, David, and Garry D. Bruton. 2006. "Venture Capital in Emerging Economies: Networks and Institutional Change." *Entrepreneurship: Theory and Practice* 30 (2): 299–320.

Ahmadjian, Christina L. 2006. "Japanese Business Groups: Continuity in the Face of Change." In *Business Groups in East Asia*, ed. Sea-Jin Chang, 29–51. New York: Oxford University Press.

Almeida, Rita, and Ana Margarida Fernandes. 2008. "Openness and Technological Innovations in Developing Countries: Evidence from Firm-Level Surveys." *Journal of Development Studies* 44 (5): 701–27.

Altonji, Joseph G., Prashant Bharadwaj, and Fabian Lange. 2008. "Changes in Characteristics of American Youth: Implications for Adult Outcomes." NBER Working Paper 13883, National Bureau of Economic Research, Cambridge, MA.

Alvarez, Roberto. 2007. "Explaining Export Success: Firm Characteristics and Spillover Effects." *World Development* 35 (3): 377–93.

Amiti, Mary, and Caroline Freund. 2007. "An Anatomy of China's Export Growth." Paper presented at the International Monetary Fund Conference on Global Implications of China's Trade, Investment, and Growth, Washington, DC, April 6.

Amsden, Alice H. 1989. *Asia's Next Giant: South Korea and Late Industrialization.* New York and Oxford, U.K.: Oxford University Press.

Anderson, Chris. 2006. *The Long Tail: Why the Future of Business Is Selling Less of More.* New York: Hyperion Books.

Archibugi, Daniele, and Alberto Coco. 2005. "Measuring Technological Capabilities at the Country Level: A Survey and a Menu for Choice." *Research Policy* 34 (2): 175–94.

Ariffin, Norlela, and Paulo N. Figueiredo. 2004. "Internationalization of Innovation Capabilities: Counter-evidence from the Electronics Industry in Malaysia and Brazil." *Oxford Development Studies* 2 (4): 559–83.

Aswicahyono, Haryo, M. Chatib Basri, and Hal Hill. 2000. "How Not to Industrialise? Indonesia's Automotive Industry." *Bulletin of Indonesian Economic Studies* 36 (1): 209–41.

Atkinson, Robert D. 2007. "The Case for a National Broadband Policy." Information Technology and Innovation Foundation, Washington, DC.

Au, Chung-Chung, and J. Vernon Henderson. 2006a. "Are Chinese Cities Too Small?" *Review of Economic Studies* 73 (2): 549–76.

———. 2006b. "How Immigration Restrictions Limit Agglomeration and Productivity in China." *Journal of Development Economics* 80 (2): 350–88.

Avnimelech, Gil, and Morris Teubal. 2006. "Creating Venture Capital Industries That Co-evolve with High Tech: Insights from an Extended Industry Life Cycle Perspective of the Israeli Experience." *Research Policy* 35 (10): 1477–98.

Aw, Bee Yan, Mark J. Roberts, and Tor Winston. 2005. "The Complementary Role of Exports and R&D Investments as Sources of Productivity Growth." NBER Working Paper 11774, National Bureau of Economic Research, Cambridge, MA.

Balisacan, Arsenio M., and Hal Hill. 2003. *The Philippine Economy: Development, Policies, and Challenges*. New York: Oxford University Press.

Barro, Robert J., and Jong-Wee Lee. 2000. "International Data on Educational Attainment: Updates and Implications." CID Working Paper 042, Center for International Development, Harvard University, Cambridge, MA.

Basiron, Yusof. 2005. "Biofuel: An Alternative Fuel in the Malaysian Scenario." Malaysian Palm Oil Board, Kuala Lumpur.

Basiron, Yusof, N. Balu, and D. Chandramohan. 2004. "Palm Oil: The Driving Force of World Oils and Fats Economy." *Oil Palm Industry Economic Journal* 4 (1): 1–10.

Becker, Charles M., Jeffrey G. Williamson, and Edwin S. Mills. 1992. *Indian Urbanization and Economic Growth since 1960*. Baltimore, MD: Johns Hopkins University Press.

Beijing Municipal Bureau of Statistics and Beijing General Team of Investigation under the National Bureau of Statistics. 2005. *Beijing Statistical Yearbook 2005*. Beijing: China Statistics Press.

Belleflamme, Paul, Pierre Picard, and Jacques-François Thisse. 2000. "An Economic Theory of Regional Clusters." *Journal of Urban Economics* 48 (1): 158–84.

Berger, Suzanne, and Richard Lester. 1997. *Made by Hong Kong*. New York: Oxford University Press.

Bernard, Andrew B., and J. Bradford Jensen. 1999. "Exporting and Productivity." NBER Working Paper 7135, National Bureau of Economic Research, Cambridge, MA.

———. 2001. "Who Dies? International Trade, Market Structure, and Industrial Restructuring." CES Working Paper 01–04, Center for Economic Studies, U.S. Bureau of the Census, Suitland, MD.

Bernard, Andrew B., J. Bradford Jensen, Stephen J. Redding, and Peter K. Schott. 2007. "Firms in International Trade." *Journal of Economic Perspectives* 21 (3): 105–30.

Bernard, Andrew B., J. Bradford Jensen, and Peter K. Schott. 2003. "Falling Trade Costs, Heterogeneous Firms, and Industry Dynamics." NBER Working Paper 9639, National Bureau of Economic Research, Cambridge, MA.

Bernardo, Romeo L., and Marie-Christine G. Tang. 2008. "The Political Economy of Reform during the Ramos Administration (1992–1998)." Commission on Growth and Development Working Paper 39, World Bank, Washington, DC.

Bertaud, Alain, Douglas Webster, Cai Jianming, Yang Zhenshan, and Andrew Gulbrandson. 2007. "Urban Land Use Efficiency in China: Issues and Policy Recommendations." World Bank, Washington, DC.

Bettencourt, Luís M. A., José Lobo, Dirk Helbing, Christian Kühnert, and Geoffrey B. West. 2007. "Growth, Innovation, Scaling, and the Pace of Life in Cities." *Proceedings of the National Academy of Science* 104 (17): 7301–06.

Bettencourt, Luís M. A., José Lobo, and Deborah Strumsky. 2007. "Invention in the City: Increasing Returns to Patenting as a Scaling Function of Metropolitan Size." *Research Policy* 36 (1): 107–20.

bin Abdul Hamid, Azhar. 2009. "Palm Oil Industry in Malaysia." Presented at the Third South-South Study Visit to Singapore: Skills and Knowledge for Sustained Development in Africa, Singapore, June 24.

Blalock, Garrick, and Paul Gertler. 2003. "Technology Acquisition in Indonesian Manufacturing: The Effect of Foreign Direct Investment." Background paper prepared for the Innovative East Asia Study, World Bank, Washington, DC.

Blalock, Garrick, and Francisco M. Veloso. 2007. "Imports, Productivity Growth, and Supply Chain Learning." *World Development* 35 (7): 1134–51.

Blonigen, Bruce, and Alyson Ma. 2007. "Please Pass the Catch-Up: The Relative Performance of Chinese and Foreign Firms in Chinese Exports." NBER Working Paper 13376, National Bureau of Economic Research, Cambridge, MA.

Bloom, Nicholas, Raffaella Sadun, and John Van Reenen. 2007. "Americans Do I.T. Better: U.S. Multinationals and the Productivity Miracle." CEPR Discussion Paper 6291, Centre for Economic Policy Research, London.

Bosworth, Barry P. 2005. "Economic Growth in Thailand: 1994–2002." World Bank, Washington, DC.

Bosworth, Barry P., and Susan M. Collins. 2007. "Accounting for Growth: Comparing China and India." NBER Working Paper 12943, National Bureau of Economic Research, Cambridge, MA.

Bozeman, Barry. 2000. "Technology Transfer and Public Policy: A Review of Research and Theory." *Research Policy* 29 (4–5): 627–55.

Brenton, Paul, and Richard Newfarmer. 2007. "Watching More than the Discovery Channel: Export Cycles and Diversification in Development." Policy Research Working Paper 4302, World Bank, Washington, DC.

Brenton, Paul, Richard Newfarmer, and Peter Walkenhorst. 2008. "Avenues for Export Diversification: Issues for Low-Income Countries." Commission on Growth and Development Working Paper 47, World Bank, Washington, DC.

Brock, William, Ray Marshall, and Marc Tucker. 2009. "10 Steps to World-Class Schools." *Washington Post*, May 30.

Brockmeier, Martina. 2001. "A Graphical Exposition of the GTAP Model." GTAP Technical Paper 8, Center for Global Trade Analysis, Purdue University, West Lafayette, IN.

Brunner, Hans-Peter, and Massimiliano Cali. 2006. "The Dynamics of Manufacturing Competitiveness in South Asia: An Analysis through Export Data." *Journal of Asian Economics* 17 (4): 557–82.

Brynjolfsson, Erik, and Lorin M. Hitt. 2003. "Computing Productivity: Firm-Level Evidence." *Review of Economics and Statistics* 85 (4): 793–808.

Buenstorf, Guido, and Steven Klepper. 2009. "Heritage and Agglomeration: The Akron Tyre Cluster Revisited." *Economic Journal* 119 (537): 705–33.

"Bugs in the Tank." 2008. *Economist*, May 2.

Bunnell, Tim. 2003. *Malaysia, Modernity, and the Multimedia Corridor: A Critical Geography of Intelligent Landscapes*. New York: RoutledgeCurzon.

Bunnell, Tim, Paul A. Barter, and Sirat Morshidi. 2002. "Kuala Lumpur Metropolitan Area: A Globalizing City-Region." *Cities* 19 (5): 357–70.

Burton, John. 2009a. "Malaysia Expects Little Recovery until 2011. *Financial Times*, March 11.

———. 2009b. "Najib Looks to Be Radically Different." *Financial Times*, July 1.

———. 2009c. "Surprise $16bn Stimulus to Boost Growth." *Financial Times*, March 11.

Butler, Rhett A. 2006. "Why Is Oil Palm Replacing Tropical Rainforests?" Mongabay.com. http://news.mongabay.com/2006/0425-oil_palm.html.

———. 2007. "The Impact of Palm Oil on the Environment." Mongabay.com. http://www.mongabay.com/borneo/borneo_oil_palm.html.

Bwalya, Samuel Mulenga. 2006. "Foreign Direct Investment and Technology Spillovers: Evidence from Panel Data Analysis of Manufacturing Firms in Zambia." *Journal of Development Economics* 81 (2): 514–26.

"Capturing Talent: Asia's Skills Shortage." 2007. *Economist*, August 18.

Carey, John. 2009. "The Biofuel Bubble." *BusinessWeek*, April 27.

Carlino, Gerald A., Satyajit Chatterjee, and Robert M. Hunt. 2007. "Urban Density and the Rate of Invention." *Journal of Urban Economics* 61 (3): 389–419.

Chandra, Vandana, Jessica Boccardo, and Israel Osorio. 2007. "Export Diversification and Competitiveness in Developing Countries." World Bank, Washington, DC.

Chang, Sea-Jin. 2006a. "Introduction: Business Groups in East Asia." In *Business Groups in East Asia*, ed. Sea-Jin Chang, 1–26. New York: Oxford University Press.

———. 2006b. "Korean Business Groups: The Financial Crisis and the Restructuring of Chaebols." In *Business Groups in East Asia*, ed. Sea-Jin Chang, 52–69. New York: Oxford University Press.

Choo, Yuen May, Ma Ah Ngan, Cheah Kien Yoo, Rusnani Abdul Majid, and Andrew Yap Kian Chung. 2005. "Palm Diesel: Green and Renewable Fuel from Palm Oil." Malaysian Palm Oil Board, Kuala Lumpur.

Chung, Young-Lob. 2007. *South Korea in the Fast Lane*. Oxford, U.K.: Oxford University Press.

Clerides, Sofronis K., Saul Lach, and James R. Tybout. 1998. "Is Learning by Exporting Important? Micro-dynamic Evidence from Colombia, Mexico, and Morocco." *Quarterly Journal of Economics* 113 (3): 903–47.

Corden, W. Max. 2007. "The Asian Crisis: A Perspective after Ten Years." *Asian-Pacific Economic Literature* 21 (2): 1–12.
"Corporate: EngTek Back on Growth Path." 2007. *Edge Malaysia*, December 3.
Coulibaly, Souleymane, Uwe Deichmann, and Somik Lall. 2007. "Urbanization and Productivity: Evidence from Turkish Provinces over the Period 1980–2000." Policy Research Working Paper 4327, World Bank, Washington, DC.
Coxhead, Ian, and Muqun Li. 2008. "Prospects for Skills-Based Export Growth in a Labour-Abundant, Resource-Rich Developing Economy." *Bulletin of Indonesian Economic Studies* 44 (2): 209–38.
Crafts, Nicholas. 2005. "Interpreting Ireland's Economic Growth." Background paper for *Industrial Development Report 2005*, United Nations Industrial Development Organization, Vienna.
Cui, Li, and Murtaza Syed. 2007. "The Shifting Structure of China's Trade and Production." IMF Working Paper 07/214, International Monetary Fund, Washington, DC.
Dahl, Michael S., Christian Ø. R. Pedersen, and Bent Dalum. 2005. "Entrepreneurial Founder Effects in the Growth of Regional Clusters: How Early Success Is a Key Determinant." DRUID Working Paper 05-18, Danish Research Unit for Industrial Dynamics, Copenhagen.
Degroof, Jean-Jacques, and Edward B. Roberts. 2004. "Overcoming Weak Entrepreneurial Infrastructures for Academic Spin-Off Ventures." MIT-IPC Working paper 04-005, Industrial Performance Center, Massachusetts Institute of Technology, Cambridge, MA.
Deichmann, Uwe, Kai Kaiser, Somik Lall, and Zmarak Shalizi. 2005. "Agglomeration, Transport, and Regional Development in Indonesia." Policy Research Working Paper 3477, World Bank, Washington, DC.
De Loecker, Jan. 2007. "Do Exports Generate Higher Productivity? Evidence from Slovenia." *Journal of International Economics* 73 (1): 69–98.
DeLong, J. Bradford, and Lawrence H. Summers. 1991. "Equipment Investment and Economic Growth." *Quarterly Journal of Economics* 106 (2): 445–502.
———. 1993. "How Strongly Do Developing Economies Benefit from Equipment Investment?" *Journal of Monetary Economics* 33 (2): 395–415.
Department of Statistics. 2003. *Labour Force Survey Report 2002*. Putrajaya, Malaysia: Department of Statistics.
———. 2007. *Labour Force Survey Report 2006*. Putrajaya, Malaysia: Department of Statistics.
Department of Trade and Industry, U.K. 2006. *The 2006 R&D Scoreboard*. London: Department of Trade and Industry.
———. 2007. *The 2007 R&D Scoreboard*. London: Department of Trade and Industry.
———. 2008. *The 2008 R&D Scoreboard*. London: Department of Trade and Industry.
Desmet, Klaus, and Esteban Rossi-Hansberg. 2007. "Spatial Growth of Employment in 'Young' Services Sector Mirroring That of Early 20th Century Manufacturing." VoxEU.org, Centre for Economic Policy Research, London. http://www.voxeu.org/index.php?q=node/492.
Diacon, Tyler, and Thomas H. Klier. 2003. "Where the Headquarters Are: Evidence from Large Public Companies 1990–2000." Working Paper 2003–35, Federal Reserve Bank of Chicago, Chicago.

Dick, Howard W., Vincent J. H. Houben, Thomas Lindblad, and Kian Wie Thee. 2002. *The Emergence of a National Economy: An Economic History of Indonesia, 1800–2000.* Honolulu: University of Hawaii Press.

Dimaranan, Betina, Elena Ianchovichina, and Will Martin. 2007. "Competing with Giants: Who Wins, Who Loses?" In *Dancing with Giants: China, India, and the Global Economy*, ed. L. Alan Winters and Shahid Yusuf, 67–100. Washington, DC: World Bank.

Draca, Mirko, Raffaella Sadun, and John Van Reenen. 2007. "Productivity and ICTs: A Review of the Evidence." In *The Oxford Handbook of Information and Communication Technologies*, ed. Robin Mansell, Chrisanthi Avgerou, Danny Quah, and Roger Silverstone, 100–47. Oxford, U.K.: Oxford University Press.

Duranton, Gilles. 2008. "From Cities to Productivity and Growth in Developing Countries." CEPR Discussion Paper 6634, Centre for Economic Policy Research, London.

Dwyer, Michael. 2007. "As China Makes More Components, Asian Neighbors Lose Out." *International Herald Tribune*, October 23.

Economist Intelligence Unit. 1996. "Country Report: Malaysia and Brunei." Economist Intelligence Unit, London.

Edwards, Sebastian. 1998. "Openness, Productivity, and Growth: What Do We Really Know?" *Economic Journal* 108 (447): 383–98.

EIA (Energy Information Administration). 2007. *International Energy Outlook 2007.* Washington, DC: U.S. Department of Energy.

Emmerson, Donald K. 1999. *Indonesia beyond Suharto: Polity, Economy, Society, Transition.* Armonk, NY: East Gate Book.

"Engineering Research at UTeM." 2007. *Prospect Malaysia*, 70.

Enright, Michael J., Edith E. Scott, and David Dodwell. 1997. *The Hong Kong Advantage.* Hong Kong, China: Oxford University Press.

EPU (Economic Planning Unit of Malaysia). 2006. *Ninth Malaysia Plan 2006–2010.* Putrajaya, Malaysia: EPU.

Ernst, Dieter. 2003. "How Sustainable Are Benefits from Global Production Networks? Malaysia's Upgrading Prospects in the Electronics Industry." Economics Study Area Working Paper 57, East-West Center, Honolulu, HI.

Etzkowitz, Henry. 2002. *Bridging Knowledge to Commercialization: The American Way.* Purchase, NY: Science Policy Institute, State University of New York.

Fan, Peilei, and Chihiro Watanabe. 2006. "Promoting Industrial Development through Technology Policy: Lessons from Japan and China." *Technology in Society* 28 (3): 303–20.

Feenstra, Robert C., and Gary C. Hamilton. 2006. *Emergent Economies, Divergent Paths: Economic Organization and International Trade in South Korea and Taiwan.* New York: Cambridge University Press.

Feldman, Maryann P. 2001. "The Entrepreneurial Event Revisited: Firm Formation in a Regional Context." *Industrial and Corporate Change* 10 (4): 861–91.

Feldman, Maryann P., and Johanna Francis. 2003. "Fortune Favours the Prepared Region: The Case of Entrepreneurship and the Capitol Region Biotech Cluster." *European Planning Studies* 11 (7): 765–88.

Fleming, Daniel, and Henrik Søborg. 2002. "Dilemmas of a Proactive Human Resource Development Policy in Malaysia." *European Journal of Development Research* 14 (1): 145–70.

Florida, Richard. 2002. *The Rise of the Creative Class and How It's Transforming Work, Leisure, and Everyday Life*. New York: Basic Books.

———. 2005a. *Cities and the Creative Class*. London: Routledge.

———. 2005b. *The Flight of the Creative Class: The New Global Competition for Talent*. New York: HarperCollins.

Florida, Richard, Gary Gates, Brian Knudsen, and Kevin Stolarick. 2006. "The University and the Creative Economy." Creative Class Group, Portland, OR.

Frost & Sullivan. 2005. "Shared Services and Outsourcing (SSO) and Hub Potential." Frost & Sullivan, Kuala Lumpur.

Furman, Jeffrey, and Scott Stern. 2006. "Climbing Atop the Shoulders of Giants: The Impact of Institutions on Cumulative Research." NBER Working Paper 12523, National Bureau of Economic Research, Cambridge, MA.

Gallagher, Kevin P., Juan Carlos Moreno-Brid, and Roberto Porzecanski. 2008. "The Dynamism of Mexican Exports: Lost in (Chinese) Translation?" *World Development* 36 (8): 1365–80.

Gans, Joshua S., and Scott Stern. 2003. "The Product Market and the Market for 'Ideas': Commercialization Strategies for Technology Entrepreneurs." *Research Policy* 32 (2): 333–50.

Gaulier, Guillaume, Françoise Lemoine, and Deniz Ünal-Kesenci. 2007. "China's Emergence and the Reorganisation of Trade Flows in Asia." *China Economic Review* 18 (3): 209–43.

Gereffi, Gary, and Miguel Korzeniewicz. 1993. *Commodity Chains and Global Capitalism*. West Port, CT: Praeger.

Glaeser, Edward L. 2007. "Entrepreneurship and the City." NBER Working Paper 13551, National Bureau of Economic Research, Cambridge, MA.

———. 2009. "Growth: The Death and Life of Cities." In *Making Cities Work: Prospects and Policies for Urban America*, ed. Robert P. Inman, 22–62. Princeton, NJ: Princeton University Press.

Gomez, Edmund Terence. 2006. "Malaysian Business Groups: The State and Capital Development in the Post-currency Crisis Period." In *Business Groups in East Asia*, ed. Sea-Jin Chang, 119–46. New York: Oxford University Press.

Gonzales, Patrick, Juan Carlos Guzmán, Lisette Partelow, Erin Pahlke, Leslie Jocelyn, David Kastberg, and Trevor Williams. 2004. *Highlights from the Trends in International Mathematics and Science Study (TIMSS) 2003*. Washington, DC: National Center for Education Statistics.

Gonzales, Patrick, Trevor Williams, Leslie Jocelyn, Stephen Roey, David Kastberg, and Summer Brenwald. 2008. *Highlights from TIMSS 2007: Mathematics and Science Achievement of U.S. Fourth- and Eighth-Grade Students in an International Context*. Washington, DC: National Center for Education Statistics.

Gott, John. 2009. "International Trade in Services Course: Global Services Location Index." Presented at the Trade in Services Brown Bag Lunch, World Bank, Washington, DC, March 27.

Graham, Edward M. 2003. *Reforming Korea's Industrial Conglomerates*. Washington, DC: Peterson Institute.

Greenaway, David, Aruneema Mahabir, and Chris Milner. 2008. "Has China Displaced Other Asian Countries' Exports?" *China Economic Review* 19 (2): 152–69.

Guarini, Giulio, Vasco Molini, and Roberta Rabellotti. 2006. "Is Korea Catching Up? An Analysis of the Labour Productivity Growth in South Korea." *Oxford Development Studies* 34 (3): 323–39.

Guimaraes, Roberto, and Olaf Unteroberdoerster. 2006. "What's Driving Private Investment in Malaysia? Aggregate Trends and Firm-Level Evidence." IMF Working Paper 06/190, International Monetary Fund, Washington, DC.

Haddad, Mona, and Ann Harrison. 1993. "Are There Positive Spillovers from Direct Foreign Investment? Evidence from Panel Data for Morocco." *Journal of Development Economics* 42 (1): 51–74.

Hallward-Driemeier, Mary, Giuseppe Iarossi, and Kenneth L. Sokoloff. 2002. "Exports and Manufacturing Productivity in East Asia: A Comparative Analysis with Firm-Level Data." NBER Working Paper 8894, National Bureau of Economic Research, Cambridge, MA.

Haltmaier, Jane T., Shaghil Ahmed, Brahima Coulibaly, Ross Knippenberg, Sylvain Leduc, Mario Marazzi, and Beth Anne Wilson. 2007. "The Role of China in Asia: Engine, Conduit, or Steamroller?" International Finance Discussion Paper 904, Board of Governors of the Federal Reserve System, Washington, DC.

Hanani, Alberto D. 2006. "Indonesia Business Groups: The Crisis in Progress." In *Business Groups in East Asia*, ed. Sea-Jin Chang, 179–204. New York: Oxford University Press.

Hanson, Gordon H., and Raymond Robertson. 2007. "China and the Manufacturing Exports of Other Developing Countries." University of California, San Diego; San Diego, CA.

Harrigan, James, and Anthony J. Venables. 2006. "Timeliness and Agglomeration." *Journal of Urban Economics* 59 (2): 300–16.

Haskel, Jonathan E., Sonia C. Pereira, and Matthew J. Slaughter. 2007. "Does Inward Foreign Direct Investment Boost the Productivity of Domestic Firms?" *Review of Economics and Statistics* 89 (3): 482–96.

Hausmann, Ricardo, Jason Hwang, and Dani Rodrik. 2007. "What You Export Matters." *Journal of Economic Growth* 12 (1): 1–25.

Hausmann, Ricardo, and Bailey Klinger. 2006. "Structural Transformation and Patterns of Comparative Advantage in the Product Space." CID Working Paper 128, Center for International Development, Harvard University, Cambridge, MA.

———. 2008. "Growth Diagnostics in Peru." CID Working Paper 181, Center for International Development, Harvard University, Cambridge, MA.

Hayes, Robert H., and Gary P. Pisano. 1994. "Beyond World-Class: The New Manufacturing Strategy." *Harvard Business Review* (January–February): 77–86.

Heckman, James J., and Paul A. LaFontaine. 2008. "The Declining American High School Graduation Rate: Evidence, Sources, and Consequences." *NBER Reporter* 1: 3–5.

Henderson, J. Vernon. 2004. "Issues Concerning Urbanization in China." Prepared for the 11th Five-Year Plan of China, World Bank, Washington, DC.

———. 2005. "Urbanization and Growth." In *Handbook of Economic Growth*, vol. 1B, ed. Philippe Aghion and Steven N. Durlauf, 1543–91. Amsterdam: Elsevier.

Henderson, Jeffrey, and Richard Phillips. 2007. "Unintended Consequences: Social Policy, State Institutions, and the 'Stalling' of the Malaysian Industrialization Project." *Economy and Society* 36 (1): 78–102.

Hill, Hal. 1996. *Indonesian Economy since 1966: Southeast Asia's Emerging Giant*. Cambridge, U.K.: Cambridge University Press.

Hillberry, Russell, and David Hummels. 2005. "Trade Responses to Geographic Frictions: A Decomposition Using Micro-data." NBER Working Paper 11339, National Bureau of Economic Research, Cambridge, MA.

Hiratsuka, Daisuke. 2006. "Outward FDI from and Intraregional FDI in ASEAN: Trends and Drivers." Discussion Paper 77, Institute of Developing Economies, Chiba, Japan.

Hobday, Mike. 1994. "Export-Led Technology Development in the Four Dragons: The Case of Electronics." *Development and Change* 25 (2): 333–61.

Hoekman, Bernard, and Beata Smarzynska Javorcik. 2006. *Global Integration and Technology Transfer*. Washington, DC: World Bank.

Hulten, Charles R., and Anders Isaksson. 2007. "Why Development Levels Differ: The Sources of Differential Economic Growth in a Panel of High and Low Income Countries." NBER Working Paper 13469, National Bureau of Economic Research, Cambridge, MA.

Hummels, David, and Peter J. Klenow. 2005. "The Variety and Quality of a Nation's Exports." *American Economic Review* 95 (3): 704–23.

Ianchovichina, Elena, Maros Ivanic, and Will Martin. 2009. "The Growth of China and India: Implications and Policy Reform Options for Malaysia." World Bank, Washington, DC.

IEA (International Energy Agency). 2004. *Biofuels for Transport: An International Perspective*. Paris: IEA Books.

———. 2006. "IEA Global Renewable Fact Sheet 2006." IEA, Paris.

Imai, Masami. 2006. "Mixing Family Business with Politics in Thailand." *Asian Economic Journal* 20 (3): 241–56.

Imbs, Jean, and Romain Wacziarg. 2003. "Stages of Diversification." *American Economic Review* 93 (1): 63–86.

IMD (International Institute for Management and Development). 2003. *World Competitiveness Yearbook*. Lausanne, Switzerland: IMD.

———. 2004. *World Competitiveness Yearbook*. Lausanne, Switzerland: IMD.

———. 2005. *World Competitiveness Yearbook*. Lausanne, Switzerland: IMD.

———. 2006. *World Competitiveness Yearbook*. Lausanne, Switzerland: IMD.

———. 2007. *World Competitiveness Yearbook*. Lausanne, Switzerland: IMD.

———. 2008. *World Competitiveness Yearbook*. Lausanne, Switzerland: IMD.

Indergaard, Michael. 2004. *Silicon Alley: The Rise and Fall of a New Media District*. New York: Routledge.

"Indonesia/Malaysia: Palm Oil Raises Concerns." 2008. *Oxford Analytica*, April 14.

"International: Land Conversion Undermines Biofuels." 2008. *Oxford Analytica*, February 28.

Jaafar, Mohd. Zamzam, Wong Hwee Kheng, and Norhayati Kamaruddin. 2003. "Greener Energy Solutions for a Sustainable Future: Issues and Challenges for Malaysia." *Energy Policy* 31 (11): 1061–72.

Jan, Tain-Sue, and Yijen Chen. 2006. "The R&D System for Industrial Development in Taiwan." *Technological Forecasting and Social Change* 73 (5): 559–74.

Jaumotte, Florence, and Nikola Spatafora. 2007. "Asia Rising: A Sectoral Perspective." IMF Working Paper 07/130, International Monetary Fund, Washington, DC.

Johnson, Chalmers A. 1982. *MITI and the Japanese Miracle: The Growth of Industrial Policy, 1925–1975*. Palo Alto, CA: Stanford University Press.

Jones, Charles I. 2005. "Growth and Ideas." In *Handbook of Economic Growth*, vol. 1B, ed. Philippe Aghion and Steven N. Durlauf, 1064–111. Amsterdam: Elsevier.

Jordaan, Jacob A. 2005. "Determinants of FDI-Induced Externalities: New Empirical Evidence for Mexican Manufacturing Industries." *World Development* 33 (12): 2103–18.

Jorgenson, Dale W., S. Ho Mun, Jon D. Samuels, and Kevin J. Stiroh. 2007. "Industry Origins of the American Productivity Resurgence." Harvard University, Cambridge, MA.

Kahn, Matthew E. Forthcoming. "New Evidence on Trends in the Cost of Urban Agglomeration." In *The Economics of Agglomeration*, ed. Edward L. Glaeser. Chicago, IL: University of Chicago Press.

Kang, Chul-Kyu. 2005. "Development of Competition Law in Korea and Current Issues." Presented at the Second East Asia Conference on Competition Law and Policy, Bogor, Indonesia, April 3.

Kaplinsky, Raphael, and Amelia Santos Paulino. 2005. "Innovation and Competitiveness: Trends in Unit Prices in Global Trade." *Oxford Development Studies* 33 (3–4): 333–55.

Kenney, Martin. 2008. "Lessons from the Development of Silicon Valley and Its Entrepreneurial Support Network for Japan and Kyushu." In *Growing Industrial Clusters in Asia: Serendipity and Science*, ed. Shahid Yusuf, Kaoru Nabeshima, and Shoichi Yamashita, 39–66. Washington, DC: World Bank.

Kenney, Martin, Kyonghee Han, and Shoko Tanaka. 2004. "Venture Capital Industries." In *Global Change and East Asian Policy Initiatives*, ed. Shahid Yusuf, M. Anjum Altaf, and Kaoru Nabeshima, 391–427. New York: Oxford University Press.

Khalafalla, Khalid Yousif, and Alan J. Webb. 2001. "Export-Led Growth and Structural Change: Evidence from Malaysia." *Applied Economics* 33 (13): 1703–15.

Kim, Linsu. 1997. *Imitation to Innovation: The Dynamics of Korea's Technological Learning*. Cambridge, MA: Harvard Business School Press.

———. 2000. "The Dynamics of Public Policy, Corporate Strategy, and Technological Learning: Lessons from the Korean Experience." College of Business Administration, Korea University, Seoul.

Kim, Sangho, and Mazlina Shafi'i. 2009. "Factor Determinants of Total Factor Productivity Growth in Malaysian Manufacturing Industries: A Decomposition Analysis." *Asian-Pacific Economic Literature* 23 (1): 48–65.

King, Elisa B. 2007. "Making Sense of the Failure of Rapid Industrialisation in the Philippines." *Technology in Society* 29 (3): 295–306.

Kinkyo, Takuji. 2007. "Explaining Korea's Lower Investment Levels after the Crisis." *World Development* 35 (7): 1120–33.

Klinger, Bailey, and Daniel Lederman. 2006. "Diversification, Innovation, and Imitation off the Global Technological Frontier." Policy Research Working Paper 3872, World Bank, Washington, DC.

Kodama, Fumio, and Jun Suzuki. 2007. "How Japanese Companies Have Used Scientific Advances to Restructure Their Business: The Receiver-Active National System of Innovation." *World Development* 35 (6): 976–90.

Kohpaiboon, Archanun. 2006. "Foreign Direct Investment and Technology Spillover: A Cross-Industry Analysis of Thai Manufacturing." *World Development* 34 (3): 541–56.

Krazit, Tom. 2007. "Intel R&D on Slow Boat to China." *CNET News*, April 16. http://www.news.com/Intel-RD-on-slow-boat-to-China/2100-1006_3-6176204.html.

Kugler, Maurice. 2006. "Spillovers from Foreign Direct Investment: Within or between Industries?" *Journal of Development Economics* 80 (2): 444–77.

Lall, Sanjaya. 2000. "The Technological Structure and Performance of Developing Country Manufactured Exports, 1985–98." *Oxford Development Studies* 28 (3): 337–69.

Lall, Sanjaya, John Weiss, and Jinkang Zhang. 2006. "The "Sophistication" of Exports: A New Trade Measure." *World Development* 34 (2): 222–37.

Lall, Somik V., Zmarak Shalizi, and Uwe Deichmann. 2004. "Agglomeration Economies and Productivity in Indian Industry." *American Economic Review* 73 (3): 643–73.

Lam, Man Kee, Kok Tat Tan, Keat Teong Lee, and Abdul Rahman Mohamed. 2009. "Malaysian Palm Oil: Surviving the Food versus Fuel Dispute for a Sustainable Future." *Renewable and Sustainable Energy Reviews* 13 (6–7): 1456–64.

Lawrence, Robert Z., and David E. Weinstein. 1999. "Trade and Growth: Import-Led or Export-Led? Evidence from Japan and Korea." NBER Working Paper 7264, National Bureau of Economic Research, Cambridge, MA.

Lee, Cassey. 2007. "Competition Policy in Malaysia." In *Competitive Advantage and Competition Policy in Developing Countries*, ed. Paul Cook, Raul Fabella, and Cassey Lee, 183–204. Cheltenham, U.K.: Edward Elgar.

Lee, Yung Joon, and Hyoungsoo Zang. 1998. "Urbanization and Regional Productivity in Korean Manufacturing." *Urban Studies* 35 (11): 2085–99.

Lerner, Josh. 2002. "When Bureaucrats Meet Entrepreneurs: The Design of Effective 'Public Venture Capital' Programmes." *Economic Journal* 112 (477): F73–84.

Levinsohn, James, and Amil Petrin. 1999. "When Industries Become More Productive, Do Firms?" NBER Working Paper 6893, National Bureau of Economic Research, Cambridge, MA.

Lie, John. 2000. *Han Unbound: The Political Economy of South Korea*. Palo Alto, CA: Stanford University Press.

Lim, David. 1973. *Economic Growth and Development in West Malaysia 1947–1970*. New York: Oxford University Press.

Lim, Hock-Eam, Judith Rich, and Mark N. Harris. 2008. "Employment Outcomes of Graduates: The Case of Universiti Utara, Malaysia." *Asian Economic Journal* 22 (3): 321–41.

Lopez-Claros, Augusto, Michael E. Porter, Klaus Schwab, and Xavier Sala-i-Martin. 2006. *The Global Competitiveness Report 2006–2007*. New York: Palgrave Macmillan.

Lyons, Kaley, and Martin Kenney. 2007. "Report to the World Bank on the Malaysian Venture Capital Industry." World Bank, Washington, DC.

Maddison, Angus. 2006. "World Population, GDP, and Per Capita GDP, 1–2003 AD." University of Groningen, Groningen, Netherlands.

Magtibay-Ramos, Nedelyn, Gemma Estrada, and Jesus Felipe. 2008. "An Input-Out Analysis of the Philippine BPO Industry." *Asian-Pacific Economic Literature* 22 (1): 41–56.

Mahadevan, Renuka. 2007a. "New Evidence on the Export-Led Growth Nexus: A Case Study of Malaysia." *World Economy* 30 (7): 1069–83.

———. 2007b. "Perspiration versus Inspiration: Lessons from a Rapidly Developing Economy." *Journal of Asian Economics* 18 (2): 331–47.

"Malaysia: Domestic Demand Is Key to Growth Forecasts." 2007. *Oxford Analytica*, September 12.

"Malaysia: Najib Eases Investment Barriers." 2009. *Oxford Analytica*, April 28.

"Malaysia: Northern Project Poses Funding Issues." 2007. *Oxford Analytica*, August 14.

"Malaysian Carmaker Proton Keen on Foreign Tie-Ups." 2008. *International Herald Tribune*, August 28.

Malaysian Palm Oil Council. 2008. "Malaysian Palm Oil: Industry Performance 2007." *Global Oil and Fats* 5 (1): 1–8.

Mankower, Joel, Ron Pernick, and Clint Wilder. 2007. "Clean Energy Trends 2007." Clean Edge, San Francisco, CA.

Marwah, Kanta, and Akbar Tavakoli. 2004. "The Effect of Foreign Capital and Imports on Economic Growth: Further Evidence from Four Asian Countries (1970–1998)." *Journal of Asian Economics* 15 (2): 399–413.

Mathews, John A. 2007. "A Biofuels Manifesto: Why Biofuels Industry Creation Should Be Priority Number One for the World Bank and for Developing Countries." Presented at the Third International Conference on Environmental, Cultural, Economic, and Social Sustainability, Chennai, India, January 4–7.

Mathews, John A., and Dong Sung Cho. 2000. *Tiger Technology: The Creation of a Semiconductor Industry in East Asia*. Cambridge, U.K.: Cambridge University Press.

Mazumdar, Joy. 2001. "Imported Machinery and Growth in LDCs." *Journal of Development Economics* 65 (1): 209–24.

Meijers, Evert. 2005. "Polycentric Urban Regions and the Quest for Synergy: Is a Network of Cities More than the Sum of the Parts?" *Urban Studies* 42 (4): 765–81.

Mellander, Charlotta, and Richard Florida. 2006. "The Creative Class or Human Capital?" CESIS Working Paper 79, Centre of Excellence for Science and Innovation Studies, Stockholm.

Merkerk, Rutger O., and Douglas K. R. Robinson. 2006. "Characterizing the Emergence of a Technological Field: Expectations, Agendas, and Networks in Lab-on-a-Chip Technologies." *Technology Analysis and Strategic Management* 18 (3–4): 411–28.

Michelacci, Claudio. 2002. "Low Returns in R&D Due to the Lack of Entrepreneurial Skills." CEMFI Working Paper 0201, Centre for Monetary and Financial Studies, Madrid.

MIDA (Malaysian Industrial Development Authority). 2006. "Malaysian Industrial Development Authority." MIDA, Kuala Lumpur.

Midelfart-Knarvik, Karen H., Henry G. Overman, Stephen J. Redding, and Anthony J. Venables. 2000. "The Location of European Industry." European Economy: Economic Paper 142, Commission of the European Communities, Brussels.

Mills, Edwin S., and Charles M. Becker. 1986. *Studies in Indian Urban Development*. New York: Oxford University Press.

Ministry of Science, Technology, and Innovation. 2004. *Science and Technology Knowledge Productivity in Malaysia Bibliometric Study 2003*. Kuala Lumpur: Ministry of Science, Technology, and Innovation.

Mody, Ashoka, and Kamil Yilmaz. 2002. "Imported Machinery for Export Competitiveness." *World Bank Economic Review* 16 (1): 23–48.

Montobbio, Fabio, and Francesco Rampa. 2005. "The Impact of Technology and Structural Change on Export Performance in Nine Developing Countries." *World Development* 33 (4): 527–47.

Moore, Gordon, and Kevin Davis. 2004. "Learning the Silicon Valley Way." In *Building High-Tech Clusters: Silicon Valley and Beyond*, ed. Timothy Bresnahan and Alfonso Gambardella, 7–39. Cambridge, U.K.: Cambridge University Press.

Moretti, Enrico. 2002. "Human Capital Spillovers in Manufacturing: Evidence from Plant-Level Production Functions." NBER Working Paper 9316, National Bureau of Economic Research, Cambridge, MA.

Morris-Suzuki, Tessa. 1994. *The Technological Transformation of Japan: From the Seventeenth to the Twenty-First Century*. Cambridge, U.K.: Cambridge University Press.

MPOB (Malaysian Palm Oil Board). 2007. *Malaysian Oil Palm Statistics 2006*. Kuala Lumpur: Ministry of Plantation Industries and Commodities.

Mullis, Ina V. S., Michael O. Martin, Eugenio J. Gonzalez, Kelvin D. Gregory, Robert A. Garden, Kathleen M. O'Connor, Steven J. Chrostowski, and Teresa A. Smith. 2000. *TIMSS 1999 International Mathematics Report: Findings from IEA's Repeat of the Third International Mathematics and Science Study at the Eighth Grade*. Washington, DC: National Center for Education Statistics.

Nabeshima, Kaoru. 2004. "Technology Transfer in East Asia: A Survey." In *Global Production Networking and Technological Change in East Asia*, ed. Shahid Yusuf, M. Anjum Altaf, and Kaoru Nabeshima, 395–434. New York: Oxford University Press.

Nambisan, Satish, and Mohanbir Sawhney. 2007. "A Buyer's Guide to the Innovation Bazaar." *Harvard Business Review* (June): 109–18.

Navaretti, Giorgio Barba, and Isidro Soloaga. 2001. "Weightless Machines and Costless Knowledge: An Empirical Analysis of Trade and Technology Diffusion." Policy Research Working Paper 2598, World Bank, Washington, DC.

Netto, Anil. 2009. "Malaysia Wakes Up to Crisis." *Asia Times*, March 17.

Nexus Associates. 2005. "Improving the Delivery System and Change Management of Skills Training Institutes." Nexus Associates, Belmont, MA.

Ng, Thiam Hee. 2006. "Foreign Direct Investment and Productivity: Evidence from the East Asian Economies." Staff Working Paper 03/2006, United Nations Industrial Development Organization, Vienna.

Nijkamp, Peter. 2003. "Entrepreneurship in a Modern Network Economy." *Regional Studies* 37 (4): 395–405.

Noor, Halim Mohd., Roger Clarke, and Nigel Driffield. 2002. "Multinational Enterprises and Technological Effort by Local Firms: A Case Study of the Malaysian Electronics and Electrical Industry." *Journal of Development Studies* 38 (6): 129–41.

Nuttall, Chris. 2007. "Silicon Valley's Founding Fathers." *Financial Times*, October 31.

OECD (Organisation for Economic Co-operation and Development). 2006. *Competitive Cities in the Global Economy*. Paris: OECD.

———. 2007. *Main Science and Technology Indicators 2007*. Release 02. Paris: OECD.

O'Mara, Margaret Pugh. 2005. *Cities of Knowledge: Cold War Science and the Search for the Next Silicon Valley*. Princeton, NJ: Princeton University Press.

Orlando, Michael J. 2004. "Measuring Spillovers from Industrial R&D: On the Importance of Geographic and Technological Proximity." *RAND Journal of Economics* 35 (4): 777–86.

Overman, Henry G., Patricia Rice, and Anthony J. Venables. 2007. "Economic Linkages across Space." CEP Discussion Paper 805, Centre for Economic Performance, London School of Economics, London.

Overman, Henry G., and Anthony J. Venables. 2005. "Cities in the Developing World." CEP Discussion Paper 0695, Centre for Economic Performance, London School of Economics, London.

Palmade, Vincent. 2005. "Industry Level Analysis: The Way to Identify the Binding Constraints to Economic Growth." Policy Research Working Paper 3551, World Bank, Washington, DC.

Petersburg, Ofer. 2006. "Intel's Largest Research Center Planned in Haifa." *Ynetnews.com*, February 26. http://www.ynetnews.com/articles/0,7340,L-3221360,00.html.

Phongpaichit, Pasuk, and Chris Baker. 2002. *Thailand: Economy and Politics*. Oxford, U.K.: Oxford University Press.

———. 2008. *Thai Capital after the 1997 Crisis*. Bangkok: Silkworm Books.

Pilling, David. 2009. "Fatal Flaws That Wrecked Thailand's Promise." *Financial Times*, April 30.

Pinheiro-Machado, Rita, and P. L. Oliveira. 2004. "A Comparative Study of Patenting Activity in U.S. and Brazilian Scientific Institutions." *Scientometrics* 61 (3): 323–38.

Polsiri, Piruna, and Yupana Wiwattanakantang. 2006. "Thai Business Groups: Crisis and Restructuring." In *Business Groups in East Asia*, ed. Sea-Jin Chang, 147–78. New York: Oxford University Press.

Porter, Michael, Klaus Schwab, and Xavier Sala-i-Martin. 2007. *The Global Competitiveness Report 2007–2008*. New York: Palgrave McMillan.

Prahalad, C. K., and M. S. Krishnan. 2008. *The New Age of Innovation: Driving Co-created Value through Global Networks*. New York: McGraw-Hill.

"Proton Bomb." 2004. *Economist*, May 6.

PSDC (Penang Skills Development Centre). 2007. "Technology Roadmap for the Electrical and Electronics Industry of Penang." Penang, Malaysia: PSDC.

Ramasamy, Bala, Anita Chakrabarty, and Madeleine Cheah. 2004. "Malaysia's Leap into the Future: An Evaluation of the Multimedia Super Corridor." *Technovation* 24 (11): 871–83.

Ramasamy, Bala, and Matthew Yeung. 2007. "Malaysia: Trade Policy Review 2006." *World Economy* 30 (8): 1193–208.

Ran, Jimmy, Jan P. Voon, and Guangzhong Li. 2007. "How Does FDI Affect China? Evidence from Industries and Provinces." *Journal of Comparative Economics* 35 (4): 774–99.

Rasiah, Rajah. 2003. "Foreign Ownership, Technology, and Electronics Exports from Malaysia and Thailand." *Journal of Asian Economics* 14 (5): 785–811.

———. 2004. "Exports and Technological Capabilities: A Study of Foreign and Local Firms in the Electronics Industry in Malaysia, the Philippines, and Thailand." *European Journal of Development Research* 16 (3): 587–623.

———. 2007. "The Systemic Quad: Technological Capabilities and Economic Performance of Computer and Component Firms in Penang and Johor." *International Journal of Technological Learing, Innovation, and Development* 1 (2): 179–203.

———. 2008. "Malaysian Electric-Electronics Industry: Composition, Problems, and Recommendations." World Bank, Washington, DC.

Ravenhill, John, and Richard F. Doner. 2007. "Malaysia's Automobile Industry: Confronting the Liberalization Challenge." World Bank, Washington, DC.

Rawski, Evelyn. 1979. *Education and Popular Literacy in Ch'ing China*. Ann Arbor: University of Michigan Press.

Ready, Douglas A., and Jay A. Conger. 2007. "Make Your Company a Talent Factory." *Harvard Business Review* (June): 69–77.

Reinhardt, Nola. 2000. "Back to Basics in Malaysia and Thailand: The Role of Resource-Based Exports in Their Export-Led Growth." *World Development* 28 (1): 57–77.

Rice, Patricia, Anthony J. Venables, and Eleonora Patacchini. 2006. "Spatial Determinants of Productivity: Analysis for the Regions of Great Britain." *Regional Science and Urban Economics* 36 (6): 727–52.

Roberts, Mark J., and James R. Tybout. 1997. "Producer Turnover and Productivity Growth in Developing Countries." *World Bank Research Observer* 12 (1): 1–18.

Rosenthal, Stuart S., and William C. Strange. 2004. "Evidence on the Nature and Sources of Agglomeration Economics." In *Handbook of Regional and Urban Economics*, vol. 4, ed. J. Vernon Henderson and Jean-François Thisse, 2119–71. Amsterdam: North-Holland.

———. Forthcoming. "Small Establishments/Big Effects: Agglomeration, Industrial Organization and Entrepreneurship," In *The Economics of Agglomeration*, ed. Edward L. Glaeser. Chicago, IL: University of Chicago Press.

Rothkopf, Garten. 2007. "A Blueprint for Green Energy in the Americas." Inter-American Development Bank, Washington, DC.

Routti, Jorma. 2003. "Research and Innovation in Finland: Transformation into a Knowledge Economy." Presented at the Knowledge Economy Forum II, Helsinki, March 26.

Sabel, Charles F., and Annalee Saxenian. 2008. "Report on Finland." Government of Finland, Helsinki.

Scharlemann, Jörn P. W., and William F. Laurance. 2008. "How Green Are Biofuels?" *Science* 319 (5859): 43–44.

Schmitz, Hubert. 2005. *Value Chain Analysis for Policy Makers and Practitioners*. Geneva: International Labour Organization.

Schot, Johan, and Frank W. Geels. 2007. "Niches in Evolutionary Theories of Technical Changes: A Critical Survey of the Literature." *Journal of Evolutionary Economics* 17 (5): 605–22.

Schott, Peter K. 2001. "Do Rich and Poor Countries Specialize in a Different Mix of Goods? Evidence from Product-Level U.S. Trade Data." NBER Working Paper 8492, National Bureau of Economic Research, Cambridge, MA.

———. 2006. "The Relative Sophistication of Chinese Exports." NBER Working Paper 12173, National Bureau of Economic Research, Cambridge, MA.

Scott, Allen J. 1998. *Regions and the World Economy: The Coming Shape of Global Production, Competition, and Political Order.* New York: Oxford University Press.

Shapira, Philip, Jan Youtie, K. Yogeesvaran, and Zakiah Jaafar. 2006. "Knowledge Economy Measurement: Methods, Results, and Insights from the Malaysian Knowledge Content Study." *Research Policy* 35 (10): 1522–37.

Sharma, Y. C., and B. Singh. 2009. "Development of Biodiesel: Current Scenario." *Renewable and Sustainable Energy Reviews* 13 (6–7): 1646–51.

Sjöholm, Fredrik. 1999. "Do Foreign Contacts Enable Firms to Become Exporters?" Working Paper in Economics and Finance 326, Stockholm School of Economics, Stockholm.

Smilor, Raymond, Niall O'Donnell, Gregory Stein, and Robert Welborn. 2005. "The Research University and the Development of High Technology Centers in the U.S." Presented at the Conference on University-Industry Linkages in Metropolitan Areas in Asia, World Bank, Washington, DC, November 17.

Smith, Richard D., Rupa Chanda, and Viroj Tangcharoensathien. 2009. "Trade in Health-Related Services." *Lancet* 373 (9663): 593–601.

Soubbotina, Tatyana P. 2006. "Generic Models of Technological Learning by Developing Countries." World Bank, Washington, DC.

"South Korea: FDI Key to Unlocking Services Potential." 2004. *Oxford Analytica*, September 7.

Spence, Michael, Patricia Clarke Annez, and Robert M. Buckley, eds. 2009. *Urbanization and Growth.* Washington, DC: World Bank.

State Government of Penang. 2001. "The Second Penang Strategic Development Plan, 2001–2010." State Government of Penang, Penang, Malaysia.

Stiglitz, Joseph E. 2007. "Fifty Years of Independence: Reflections on Malaysia's Past and Future." Presented at the Khazanah Global Lectures, Kuala Lumpur, August 29.

Stone, Susan, and Anna Strutt. 2009. "Transport Infrastructure and Trade Facilitation in the Greater Mekong Subregion." ADBI Working Paper 130, Asian Development Bank Institute, Tokyo.

Straub, Stephane. 2007. "Infrastructure: Recent Advances and Research Challenges." World Bank, Washington, DC.

Studwell, Joe. 2007. *Asian Godfathers: Money and Power in Hong Kong and Southeast Asia.* Berkeley, CA: Atlantic Monthly Press.

Sumathi, S., S. P. Chai, and A. R. Mohamed. 2008. "Utilization of Oil Palm as a Source of Renewable Energy in Malaysia." *Renewable and Sustainable Energy Reviews* 12 (9): 2404–21.

Takahashi, Dean. 2007. "Fairchild's Failures Led to Gigantic Successes: Pioneering Tech Firms Spawned Many Spinoffs." *San Jose Mercury News*, October 4.

Tan, David. 2007. "Eng Teknologi Counts on China Operations." *Star*, October 8.

———. 2009. "High-End R&D Work Migrating to Penang." *Star*, March 2.

Tan, Deyi, and Chetan Ahya. 2009. "Malaysia Economics: GDP Downgrade: Facing a Triple Whammy." Morgan Stanley, New York.

Tan, Hong W., and Indermit S. Gill. 2000. "Malaysia," In *Vocational Education & Training Reform: Matching Skills to Markets and Budgets*, ed. Indermit S. Gill, Fred Fluitman, and Amit Dar, 218–60. Washington, DC: World Bank.

Tham, Siew-Yean. 2004. "Investment Regime: Malaysia." Universiti Kebangsaan Malaysia, Kuala Lumpur.

Thursby, Jerry, and Marie Thursby. 2007. "Here or There? A Survey of Factors in Multinational R&D Location." Report to the Government-University-Industry Research Roundtable, National Academies Press, Washington, DC.

Tongzon, Jose Lelis. 2007. "Malaysia and Its Future Industrial Directions." Inha University, Incheon, Republic of Korea.

Udomsaph, Charles C., and Albert Zeufack. 2006. "Skill Shortages and Mismatch in Malaysia and Thailand: Evidence from Linked Employer-Employee Data." World Bank, Washington, DC.

UNCTAD (United Nations Conference on Trade and Development). 2009. *World Investment Prospects Survey 2009–2011*. Vienna: UNCTAD.

Unger, Danny. 1998. *Building Social Capital in Thailand: Fibers, Finance, and Infrastructure*. Cambridge, U.K.: Cambridge University Press.

Venables, Anthony J., and Patricia Rice. 2005. "Spatial Determinants of Productivity: Analysis for the Regions of Great Britain." CEPR Discussion Paper 0642, Centre for Economic Policy Research, London.

Vyakarnam, Shailendra. 2007. "Understanding the Cambridge Phenomenon: Importance for Enterprise Development." Presented at the World Bank, Washington, DC, July 25.

Wade, Robert. 2003. *Governing the Market: Economic Theory and the Role of Government in East Asian Industrialization*. Princeton, NJ: Princeton University Press.

Wang, Zhi, and Shang-Jin Wei. 2007. "The Rising Sophistication of China's Exports: Assessing the Roles of Processing Trade, Foreign Invested Firms, Human Capital, and Government Policies." Columbia University, New York.

Warr, Peter. 2007. "Long-Term Economic Performance in Thailand." *ASEAN Economic Bulletin* 24 (1): 138–63.

Weber, Max. 2002. *The Protestant Ethic and "the Spirit of Capitalism."* New York: Penguin Books.

Whalley, John, and Xian Xin. 2006. "China's FDI and Non-FDI Economies and the Sustainability of Future High Chinese Growth." NBER Working Paper 12249, National Bureau of Economic Research, Cambridge, MA.

Wong, Poh-Kam. 2008. "Coping with Globalization of Production Networks and Digital Convergence: The Challenge of ICT Cluster Development in Singapore." In *Growing Industrial Clusters in Asia: Serendipity and Science*, ed. Shahid Yusuf, Kaoru Nabeshima, and Shoichi Yamashita, 91–146. Washington, DC: World Bank.

Wong, Poh-Kam, Yii Tan Chang, and Ming Yu Cheng. 2007. "Indigenous Innovative Companies in Malaysia." World Bank, Washington, DC.

Woo, Wing Thye, Bruce Glassburner, and Anwar Nasution. 1994. *Macroeconomic Policies, Crises, and Long-Term Growth in Indonesia, 1965–1990*. Washington, DC: World Bank.

World Bank. 2005a. *Doing Business in 2006*. Washington, DC: World Bank.

———. 2005b. "Malaysia: Firm Competitiveness, Investment Climate, and Growth." World Bank, Washington, DC.

———. 2006a. "Assessing the Effectiveness and Impact of Fiscal Incentives in Malaysia." World Bank, Washington, DC.

———. 2006b. *Doing Business in 2007*. Washington, DC: World Bank.

———. 2007a. *China's Information Revolution: Managing the Economic and Social Transformation*. Washington, DC: World Bank.

———. 2007b. *Doing Business in 2008*. Washington, DC: World Bank.

———. 2007c. "Malaysia and the Knowledge Economy: Building a World-Class Higher Education System." World Bank, Washington, DC.

———. 2008a. "Competitive Cities in Latin America: Discussing the Role of Investment Climate, Institutions and Infrastructure." World Bank, Washington, DC.

———. 2008b. *Doing Business in 2009*. Washington, DC: World Bank.

———. 2008c. *Global Economic Prospects*. Washington, DC: World Bank.

———. 2008d. *Towards a Knowledge Economy in Thailand*. Bangkok: World Bank.

———. 2008e. *World Development Report 2008: Agriculture for Development*. Washington, DC: World Bank.

———. 2009. *World Development Report 2009: Reshaping Economic Geography*. Washington, DC: World Bank.

World Trade Organization. 2007. *International Trade Statistics 2007*. Geneva: World Trade Organization.

Wright, Mike, Sarika Pruthi, and Andy Lockett. 2005. "International Venture Capital Research: From Cross-Country Comparisons to Crossing Borders." 7 (3): 135–65.

Xu, Bin. 2007. "Measuring China's Export Sophistication." China Europe International Business School, Shanghai, China.

Yao, Shujie, and Kailei Wei. 2007. "Economic Growth in the Presence of FDI: The Perspective of Newly Industrialising Economies." *Journal of Comparative Economics* 35 (1): 211–34.

Yusof, Zainal Aznam, and Deepak Bhattasali. 2008. "Economic Growth and Development in Malaysia: Policy Making and Leadership." Commission on Growth and Development Working Paper 27, World Bank, Washington, DC.

Yusuf, Shahid. 2007. "About Urban Mega Regions: Knowns and Unknowns." Policy Research Working Paper 4252, World Bank, Washington, DC.

———. 2008. "Can Clusters Be Made to Order?" In *Growing Industrial Clusters in Asia: Serendipity and Science*, ed. Shahid Yusuf, Kaoru Nabeshima, and Shoichi Yamashita, 1–38. Washington, DC: World Bank.

Yusuf, Shahid, M. Anjum Altaf, Barry Eichengreen, Sudarshan Gooptu, Kaoru Nabeshima, Charles Kenny, Dwight H. Perkins, and Marc Shotten. 2003. *Innovative East Asia: The Future of Growth*. New York: Oxford University Press.

Yusuf, Shahid, M. Anjum Altaf, and Kaoru Nabeshima, eds. 2004. *Global Production Networking and Technological Change in East Asia*. New York: Oxford University Press.

Yusuf, Shahid, and Kaoru Nabeshima, eds. 2007. *How Universities Promote Economic Growth*. Washington DC: World Bank.

Yusuf, Shahid, Kaoru Nabeshima, and Dwight H. Perkins. 2005. *Under New Ownership: Privatizing China's State-Owned Enterprises*. Stanford, CA: Stanford University Press.

Yusuf, Shahid, Kaoru Nabeshima, and Shoichi Yamashita, eds. 2008. *Growing Industrial Clusters in Asia: Serendipity and Science*. Washington, DC: World Bank.

Yusuf, Shahid, Shuilin Wang, and Kaoru Nabeshima. 2009. "Fiscal Policies for Innovation." In *Innovation for Development and the Role of Government: A Perspective from the East Asia and Pacific Region*, ed. Qimiao Fan, Kouqing Li, Douglas Zhihua Zeng, Yang Dong, and Runzhong Peng, 149–80. Washington, DC: World Bank.

Zakariah, Abdul Rashid, and Elyas Elameer Ahmad. 1999. "Sources of Industrial Growth Using the Factor Decomposition Approach: Malaysia, 1978–87." *Developing Economies* 37 (2): 162–96.

Zeng, Ming, and Peter Williamson. 2006. *Dragons at Your Door: How Chinese Cost Innovation Is Disrupting Global Competition*. Cambridge, MA: Harvard Business School Press.

Zhang, Chunlin. 2008. "China: Promoting Enterprise-Led Innovation." World Bank, Beijing.

Zheng, Xiao-Ping. 2007. "Economies of Network, Urban Agglomeration, and Regional Development: A Theoretical Model and Empirical Evidence." *Regional Studies* 41 (5): 559–69.

Ziderman, Adrian. 2003. *Financing Vocational Training in Sub-Saharan Africa*. Washington, DC: World Bank.

Index

Figures, notes, and tables are indicated by *f*, *n*, and *t*, respectively.

A
advanced electronic display, 107
Advanced Micro Devices (AMD), 180–81
agglomeration economies, 119–26, 123–30*t*,
    216–17. *See also* specific Malaysian cities
Agilent Technologies, 181
agriculture
    in Malaysia, 22, 23*f*, 117
    MARDI, 165–67, 166*t*, 168*t*
    palm oil sector, 6*n*11, 58, 106*t*, 113–17.
        See also Malaysian Palm Oil Board
    R&D spending on, 166, 167*t*
    in Southeast Asian tiger economies, 4, 5*t*
AKN Technology, 107
animation industry in Philippines, 7*n*16
ArCo, 28*t*
ASEAN (Association of Southeast Asian
    Nations), 49, 51, 110, 148*f*
Asia. *See also* East Asian tiger economies;
    Southeast Asian tiger economies;
    specific countries
    financial crisis of 1997–98, 3, 8, 13*n*24
    IT infrastructure in, 148*f*
Association of Southeast Asian Nations
    (ASEAN), 49, 51, 110, 148*f*
Australia, 112*n*11, 122, 146–47*t*, 217
authoritarian political regimes, 6, 7
automotive parts industry in Malaysia, 59*t*, 60,
    106*t*, 109–13, 163–65, 164–65*t*

B
back-office process outsourcing in Philippines,
    7*n*16
BF Goodrich, 141
biofuels from palm oil, 6*n*11, 58, 106*t*, 113–17
biomedical sector, 106*n*1, 181–82
Boston, cluster development in, 124, 157

Brazil, 2, 116, 160, 161*f*, 163*f*, 175*f*, 176*n*12
Bumiputra firms, 13
business efficiency, 31*t*, 32
business environment, 30, 34–35*t*
business groups, 12–13

C
Cambridge Consultants (CCL), 140–41, 144
Canada, 95
chaebols, 13*n*24
China
    automotive industry in, 110
    backward links from MNCs in, 10
    business environment, 30, 34*t*
    competition with Southeast Asian export
        economies, 8
    in competitiveness indexes, 31–33*t*
    electronics industry in, 107–8
    export of Malaysian cars to, 112*n*11
    exports from
        competitive overlap with Malaysia, 47–60,
            50*t*, 52–59*t*
        diversification of, 64–65*t*
        electronics, 43*f*, 60–62, 61–63*f*, 67, 69*t*
        as growth market, 39*t*
        leading export sectors, 45*t*
        mapping product spaces, 82, 85*f*, 225–26*t*
        MNCs as source of, 102
        plants and equipment imported to
            Malaysia, 96–97
        quality of tradable products, 73, 75, 76*t*, 77
    FDI to, 98
    imports of parts and components into, 61–62,
        61–63*f*
    innovation in
        access to finance, 150, 151–52*f*, 154*n*39,
            156, 157

269

China *(Continued)*
　educational attainment, 135*f*, 137*t*
　financial incentives for, 241–42*t*
　IRCA, 237*t*
　IT infrastructure, 145*t*, 147*t*, 148*f*, 149*t*
　large firms, 140
　licensing and technology transfers, 178–80, 179–80*f*
　patenting activities, 174*t*, 175*f*, 177*t*, 178*t*
　R&D spending, 160–61*t*, 161–62, 161–63*f*, 193*t*
　urban agglomeration economies, 120*n*5, 123–24*t*, 123*n*15
　literacy culture of, 4
　logistics services, 91, 92
　macroeconomic indicators, 29*t*
　Malaysia, Chinese-owned firms in, 13
　Malaysian economy, effects on, 199–201
　Malaysian exports to, 39, 41*f*
　production capacity, expansion of, 20
　RCA, 219*t*
　share of world electronics and machinery production, 23
　technological capabilities in, 26, 28*t*
　TFP growth in, 20, 21*t*
　value-added ratios in machinery, 27*f*
　WTO, accession to, 49
cluster development, 103, 106, 120, 124, 140–41, 142*f*
competition
　in agglomeration economies and clusters, 120*n*3
　export overlap. *See under* exports
　indexes of competitiveness, 31–33*t*, 31–34
　no Malaysian law on, 95
　openness and competitiveness in Southeast Asian tiger economies, 14
conductive polymers, 108
Confucian cultures, economic advantages of, 2*n*4
credit, availability of, 150–51, 151–52*f*
cross-shareholding, 12–13
crystal framework, 159*n*1
cultural factors associated with tiger economies, 2

D
Denmark, 141
DENSO, 181
directorships, interlocking, 12–13

diversification
　capacity for, 8–14
　of exports, 60, 62–69, 64–69*t*
doing business (business environment), 30, 34–35*t*
Doing Business Indicators, 150
domestic consumption as growth driver, 18–19, 18*t*
dynamic revealed competitiveness (DRC), 51–60, 52–54*t*, 58–59*t*

E
East Asian financial crisis of 1997–98, 3, 8, 13*n*24
East Asian tiger economies. *See also* Hong Kong, China; Korea, Republic of; Singapore; Taiwan, China
　development of, 1–2
　growth rates for
　　education, 192–93, 194–96*f*
　　GDP, 189–91, 189*t*, 190–93*f*
　　R&D spending, 192, 193*t*
　as model for Southeast Asian countries, 5
education and knowledge economy, 131–39
　access to venture capital and, 157
　comparative long-term growth statistics, 192–99, 194–98*f*
　in competitiveness indexes, 32, 33*t*, 34
　distribution of labor force by, 131*t*
　electronics industry in Malaysia and, 108–9
　English language proficiency, 131–32, 132*t*, 134–35, 138
　growth rates in Malaysia, 194*f*, 197*f*
　HRDF and training opportunities, 133–34
　innovation capacities, as means of improving, 11
　knowledge-intensive skills and industries, Malaysian efforts to foster, 25–28, 28*t*
　patenting activities of universities, 172, 173*t*
　policy incentives for, 203–8, 204*t*
　R&D institutes, tertiary, 137, 138–39*t*, 167–70, 169*t*, 171*t*–*f*
　secondary level, 134–35, 135–36*t*, 135*f*
　shortages related to, 131–33*t*
　STEM skills, importance of emphasizing, 11
　tertiary level, 131, 132, 136–38, 137*t*, 194–98*f*
　urban economies, universities in, 121–22, 125–26
electronics industry, 105–9
　China, Malaysian export of parts and components to, 61–62, 61–63*f*
　Chinese specialization in electronic exports, 43*f*, 60–62, 61–63*f*, 67, 69*t*

diversification of exports, 66–67, 67t
education and training requirements, 108–9
export market share, 42, 43f
FDI affecting, 22
increasing importance of, 23, 24t
investment trends, 62, 64t
origins in Malaysia, 17n1
patenting activities, 175, 177–78t
policy recommendations for, 209
potential versus reality in, 106–8
R&D in, 108, 165f, 167t
unit values, 69–74, 70–73f
value-added ratios in, 24–25, 27f–t
employment. *See* education and knowledge economy; labor and employment issues
Eng Teknologi, 107–8, 140
English language proficiency, 131–32, 132t, 134–35, 138
entrepreneurship, fostering, 11–12
Europe/European Union
 automotive industry in, 110
 biofuel requirements, 116n26
 food processing industry in, 117
 imports into, 37, 51, 53, 54t, 57t, 58–60, 59t, 77
 MNCs from, 7
 technology transfers to Malaysia from, 96
 urban economies in, 124
evolution of Malaysian economy, 17–35
 business environment, 30, 34–35t
 in competitiveness indexes, 31–33t, 31–34
 GDP. *See under* gross domestic product
 macroeconomic indicators, 28–30, 29t
 manufacturing sector. *See* industrialization and manufacturing
 productivity. *See* productivity issues
 sources of growth, 17–22, 18–21t
 strategy for, 17
exchange rate policy, 217
exports from Malaysia. *See also under* other specific countries
 capacity for expansion, 86–87
 changes in industrial/export mix, 8, 9t
 competitive overlap between Malaysia and comparator economies, 44–60
  with China, 47–60, 50t, 52–59t
  DRC, concept of, 51–60, 52–54t, 58–59t
  EU, exports to, 51, 53, 54t, 57t, 58–60, 59t
  Japan, exports to, 51, 53t, 54, 56t, 58–60, 59t

  with other Southeast Asian economies, 44–49, 47t, 48f
  products most affected by, 58–60
  US, exports to, 51–54, 52t, 55t, 58t
 destinations of, 77, 78t. *See also under* Europe/European Union; Japan; United States
 development of economies oriented towards, 5, 7
 diversification, 60, 62–69, 64–69t
 electronics market share, 42, 43f
 GDP, as share of, 37, 38t, 49n10, 51
 growth rates, 29t, 30, 39–40t
 growth strategy based on, 17–18, 18t, 38
 industrial change detected from, 38–43, 39t, 40–43f, 44–45t, 46f
 mapping product spaces, 81–86, 83–85f, 105, 112, 225–33t
 PRODY value, 74–77, 76–77t
 quality of tradable products, 69–77, 70–73f, 76–78t
 RCA, 77–81, 79–81t
 services, scope for expanding, 87–92, 88–89f, 90t
 specialization, 60–62, 61–62f, 208–9
 unit values, 69–74, 70–73f

F

factor inputs, 20t
Fairchild Semiconductor, 140, 144
farming. *See* agriculture
FDI. *See* foreign direct investment
Federal Land Development Agency (Malaysia), 6n11, 113
fiber optics components, 107
financial crises
 East Asian financial crisis of 1997–98, 3, 8, 13n24
 worldwide recession of 2008–2009, 187
financial investment
 access to finance in tiger economies, 150–57, 151–52f, 153t, 154f, 155t
 electronics industry, trends in, 62, 64t
 in equipment, 18n3
 FDI. *See* foreign direct investment
 GDP, gross domestic investment as percentage of, 96
 innovation, financial incentives for
  in China, 241–42t
  in Thailand, 245–48t

financial investment *(Continued)*
  Investment Incentives Act of 1968
    (Malaysia), 17
  neighboring economies' effects on
    Malaysia, 200
  policy incentives for, 206–7t
  private investment, role of, 18–21
  public investment as growth driver, 18t, 19
  in R&D. *See* research and development
  venture capital, 151–57, 153t, 154f, 155t
financial services, 90–91
Finisar, 107
Finland, 146–47t, 157, 208, 210
Florida, Richard, 121–22
flying geese model, 1n1
foreign direct investment (FDI), 98–104
  analytical use of, 15
  concerns of Southeast Asian tiger economies
    regarding, 22
  East Asian tiger economies fueled by, 1
  efforts of Malaysia to attract, 7, 25
  on global competitiveness indexes, 32
  industrial export mix in Malaysia changed by,
    8, 22, 86n24, 87
  net inflows to Southeast Asia, 98f
  passive versus active FDI-dependent learning
    countries, 159n1
  Plaza Accord (1985) affecting flow of, 8n19,
    67n17
  source economies for Malaysian FDI, 101f
  technology spillovers from, 102–4
  top 12 sectors in Malaysia, 99–100f
  urban share of, 125, 126t
France, 95
Fraunhofer institutes (Germany), 213–14
Free Trade Zone Act of 1972 (Malaysia), 17

G
GDP. *See* gross domestic product
Germany
  Fraunhofer institutes, 213–14
  growth rates for
    education, 194f
    GDP, 187, 188f, 194f
    R&D spending, 191, 193t
  IT infrastructure in, 146t, 147t
  R&D investment in, 95
global competitiveness indexes, 31–33t, 31–34
global recession of 2008, 30
Global Trade Analysis Project (GTAP),
  199, 201

government-controlled corporations (GCCs), 12
government efficiency, 31–32, 31t
government research institutes (GRIs)
  patenting activities, 172, 173t
  R&D spending by, 165–67, 167–68t,
    171t–f, 185
gross domestic product (GDP)
  analytical use of, 15
  comparative long-term growth statistics,
    187–91, 188–93f, 189t, 194–98f
  contribution of electronics, automotive parts,
    and palm oil industries to, 105, 106t
  credit-to-GDP ratio, 151f
  exports as percentage of, 37, 38t, 49n10, 51
  gross domestic investment as percentage
    of, 96
  growth rates in Malaysia, 3n6, 18t, 21n8, 30,
    189t, 191f
  macroeconomic indicators as percentage of,
    28–30, 29t
  natural resources/manufacturing balance, 4,
    5t, 17–18
  R&D spending as percentage of, 160t, 161f,
    187–88, 188–89f, 189t
  savings as share of, 28–30, 29t
  sectoral contributions to growth, 19t
  slowing growth in, 3
  tertiary education and, 194–98f
  urban share of, 123t, 125t
growth prospects, 187–201
  comparative long-term growth statistics
    education, 192–99, 194–98f
    GDP, 187–91, 188–93f, 189t, 194–98f
    R&D spending, 191–92, 193t
  effect of neighboring economies on Malaysia,
    199–201
grupos económicos in Latin America, 13n24
GTAP (Global Trade Analysis Project), 199, 201

H
Halal certification, 91
high-density storage, 107
Hong Kong, China
  educational attainment in, 134, 135–37t, 135f
  exports from, 86, 226–27t
  innovation in
    licensing and technology transfers, 179–80f
    patenting activities, 173, 174f, 175f
    R&D spending, 161f, 163f
  logistics services, 91
  product space analysis, 226–27t

service economies in, 10
tiger economy of, 1–2
Hovid, 182
HRDF (Human Resource Development Fund), Malaysia, 133–34
human capital. *See* education and knowledge economy; labor and employment issues
Human Resource Development Fund (HRDF), Malaysia, 133–34
Hybritech, 140, 142*f*

I

ICT (information and communications technology) infrastructure, 144–50, 145–47*t*, 148*f*, 149*t*, 216
IMD (International Institute for Management Development), 26, 31–32, 31*t*
IMF (International Monetary Fund), 49*n*10
imports, 93–97, 94*f*, 96*f*
India
  business environment, 30
  business houses in, 13*n*24
  exports from, 65, 75, 77, 227*t*
  innovation in, 139, 147*t*, 148*f*, 149*t*, 151–52*f*, 157, 237*t*
  IRCA, 237*t*
  Malaysian economy, effects on, 201
  product space analysis, 227*t*
  service economies in, 10, 89, 90*n*26, 210
  Tata Motors, 110
  technological capabilities in, 28*t*
  technology diffusion in, 103
  TFP growth in, 21*t*
Indonesia
  automotive industry in, 110, 111, 112*n*11
  business environment, 34*t*
  business groups in, 13*n*24
  in competitiveness indexes, 31–33*t*, 34
  development path of, 6, 8
  exports from
    competitive overlap, 43*f*, 45*t*
    diversification of, 61*f*, 64–65*f*, 65
    export structure, 9*t*
    mapping product spaces, 82, 83*f*, 86, 228*t*
    quality of products, 71, 72, 75, 76*t*, 77
  FDI to, 98
  GCCs in, 12
  growth rates for
    education, 194*f*, 197*f*
    GDP, 189*t*, 190*f*, 191, 197*f*
  innovation in
    access to finance, 151–52*f*
    educational attainment, 134, 135–36*t*, 135*f*, 147*t*
    IRCA, 238*t*
    IT infrastructure, 145*t*, 146, 147*t*, 148*f*, 149*t*, 150
    large firms, 144
    licensing and technology transfers, 179–80*f*
    patenting activities, 172, 174*t*, 175*f*, 178
    R&D spending, 160–61*t*, 161–64*f*, 162
  macroeconomic indicators, 29*t*, 30
  natural resources in, 2, 4, 5*t*, 8
  policy recommendations for, 209, 210, 216
  power shortages in, 30
  productivity effects on Malaysia, 199–200
  RCA, 220*t*
  service economies, 87, 91
  technological capabilities in, 26, 28*t*
  technology transfers, 97
  tiger economy of, 2–3
  value-added ratios in machinery, 27*f*
Industrial Coordination Act (Malaysia), 13
Industrial Technology Research Institute (Taiwan, China), 107
industrialization and manufacturing, 22–30
  analytical techniques used to study, 14–15
  automotive parts industry, 59*t*, 60, 106*t*, 109–13, 163–65, 164–65*t*
  changes in industrial/export mix, 8, 9*t*, 22–23, 23*f*, 86*n*24, 87
  development of, 2–3, 4–5, 5*t*, 6–7
  electronics. *See* electronics industry
  equipment, investment in, 18*n*3
  exports, industrial change detected from, 38–43, 39*t*, 40–43*f*, 44–45*t*, 46*f*
  GDP, manufacturing as share of, 29*t*, 30
  growth of, 17–18
  Industrial Master Plans, 106, 111, 113
  knowledge-intensive skills and technological capabilities, 25–28, 28*t*
  palm oil sector, 6*n*11, 58, 106*t*, 113–17. *See also* Malaysian Palm Oil Board
  policy incentives for, 203–8, 204–7*t*
  policy recommendations regarding, 208–14, 210–13*t*
  productivity issues, 23–30, 25–28*t*, 27*f*
  sectoral contribution to growth, 19*t*
  world production share, 23, 24*t*
information technology (IT) infrastructure, 144–50, 145–47*t*, 148*f*, 149*t*, 216

infrastructure, 30, 31–34, 31*t*, 33*t*
initial public offerings (IPOs), 155–56, 155*t*
innovation comparative advantage, index of, 183–86, 184–85*t*
innovation in Malaysia, 119–86. *See also under other specific countries*
　access to finance, 150–57, 151–52*f*, 153*t*, 154*f*, 155*t*
　capacity for innovation and diversification, 8–14, 22, 159–60
　credit, availability of, 150–51, 151–52*f*
　educational attainment. *See* education
　IT infrastructure, 144–50, 145–47*t*, 148*f*, 149*t*
　large firms, role of, 139–44, 142*f*, 143–44*t*, 214–16
　licensing and royalties, 178–80, 179–80*f*
　localization economies, 119*n*2, 120, 124
　patenting activity, 172–74*t*, 172–78, 175*f*, 176–78*t*
　policy recommendations for. *See* policy recommendations
　R&D opportunities. *See* research and development
　technology transfers, 95–97, 96*f*, 178–80
　urban industrial locations, 119–26, 123–30*t*
　venture capital, 151–57, 153*t*, 154*f*, 155*t*
innovation revealed comparative advantage (IRCA), 209, 210–13*t*, 237–39*t*
input growth, 20*t*
interlocking directorships, 12–13
Internal Security Act (Malaysia), 7*n*17
International Institute for Management Development (IMD), 26, 31–32, 31*t*
International Monetary Fund (IMF), 49*n*10
investment. *See* financial investment
Investment Incentives Act of 1968 (Malaysia), 17
IPOs (initial public offerings), 155–56, 155*t*
IRCA (innovation revealed comparative advantage), 209, 210–13*t*, 237–39*t*
Ireland, 122*n*12, 146–47*t*, 157, 159*n*1
Iskandar Development Region, Malaysia, 217*n*11
Islamic services
　finance, 90–91
　Halal certification, 91
Israel, 103, 153*n*38, 157, 175, 177–78*t*, 208
IT (information technology) infrastructure, 144–50, 145–47*t*, 148*f*, 149*t*, 216

J

Japan
　automotive industry in, 109, 112
　business environment, 34*t*
　in competitiveness indexes, 31–33*t*
　cross-shareholding in, 13*n*25
　economic and policy assistance for developing economies, 5–6
　exports from
　　competitive overlap, 47, 49
　　diversification, 64–65*t*, 65
　　electronics market share, 43*f*
　　industrial change detected from, 45*t*
　　plants and equipment imported to Malaysia, 96
　　quality of tradable products, 71–73, 75, 77
　　specialization, 61*f*, 62
　FDI from, 101*f*
　as first tiger economy, 1
　growth rates for
　　education, 194–95*f*
　　GDP, 189–91, 189*t*, 190*f*, 195*f*
　　R&D spending, 192, 193*t*
　imports into, 51, 53*t*, 54, 56*t*, 58–60, 59*t*, 77, 93
　innovation in
　　access to finance, 150, 151–52*f*
　　educational attainment, 134*f*, 135–37*t*
　　index of innovation comparative advantage, 181
　　IT infrastructure, 145–49*t*, 147
　　large firms, 139
　　licensing and technology transfers, 178, 180
　　patenting activities, 173, 175*f*, 177–78*t*
　　R&D spending, 160–61*t*, 160–62, 161–65*f*
　　receiver-active approach of firms, 159*n*1
　　urban agglomeration economies, 123–24*t*
　keiretsu in, 13*n*24
　literacy culture of, 4
　logistics services, 91
　macroeconomic indicators, 29*t*
　MNCs from, 6, 7
　Plaza Accord (1985), 8*n*19, 67*n*17
　RCA, 220*t*
　R&D investment in, 95
　technological capabilities in, 26
　value-added ratios in machinery, 27*f*
Johor, Malaysia
　education and knowledge economy in, 138–39*t*
　electronics industry in, 108, 109, 125*n*17
　FDI flows in, 100, 104, 125, 126*t*
　GDP of, 125*t*
　as innovative urban location, 124–26, 125–26*t*, 128*t*

Index **275**

population of, 125*t*
size of firms in, 141
Johor Skills Development Centre, 109

K
keiretsu, 13*n*24
Kenya, 2
Korea, Republic of
 backward links from MNCs in, 10
 business environment, 34*t*
 chaebols in, 13*n*24
 in competitiveness indexes, 31–33*t*, 34
 exports from
  competitive overlap between Malaysia and, 49
  diversification, 64–65*t*, 65, 67
  electronics market share, 43*f*
  growth rates, 39*t*
  as growth strategy, 5
  industrial change detected from, 45*t*
  mapping product spaces, 86, 229*t*
  plants and equipment imported to Malaysia, 96–97
  quality of products, 71, 72, 75, 76*t*, 77
  specialization, 61*f*, 62
 growth rates for
  education, 192, 194–95*f*
  GDP, 189*t*, 191*f*, 195*f*
  R&D spending, 192, 193*t*
 innovation in
  access to finance, 150, 151–52*f*, 156, 157
  educational attainment, 134, 135–37*t*, 135*f*
  IT infrastructure, 144, 145*t*, 146–47*t*, 149*t*
  large firms, 139, 140*n*27
  licensing and technology transfers, 19–180*f*, 178, 180
  patenting activities, 173, 174*t*, 175*f*, 177–78*t*
  R&D spending, 160–61*t*, 160–63, 161–65*f*
  urban economies, 120*n*6, 123–24*t*
 macroeconomic indicators, 29*t*
 Malaysia compared, 10, 11
 policy recommendations drawn from, 208*n*4, 217
 private investment in, 19*n*4
 RCA, 80, 221*t*
 technological capabilities in, 26, 28*t*
 tiger economy of, 1–2
 value-added ratios in machinery, 27*f*
Kuala Lumpur, Malaysia
 education and knowledge economy in, 108, 125–26, 138–39*t*

 FDI flows in, 100, 104, 126*t*
 GDP of, 125*t*
 industrialization and changing export mix in, 8
 as innovative urban location, 124–26, 125–26*t*, 127*t*, 216–17
 logistics services in, 91–92
 MSC, 149–50
 population of, 125*t*
 size of firms in, 140, 141–44, 144*t*

L
labor and employment issues
 education and skills. *See* education and knowledge economy
 expansion of workforce, 28
 productivity. *See* productivity issues
 wages in Malaysia, 24, 26*t*
large firms and innovation, 139–44, 142*f*, 143–44*t*, 214–16
Latin America and Caribbean. *See also* specific countries
 grupos económicos in, 13*n*24
 IT infrastructure in, 148*f*
 urban areas in, 122*n*13
LED technology, 107
licensing and royalties, 178–80, 179–80*f*
Linkabit, 140
localization economies, 119*n*2, 120, 124
logistics services, 91–92, 210
Lotus, 109*n*7, 112

M
macroeconomic indicators, 28–30, 29*t*
macroeconomic policy requirements, 217–18
Malacca, Malaysia
 education and knowledge economy in, 138–39*t*
 FDI flows in, 100, 126*t*
 GDP of, 125*t*
 as innovative urban location, 124–26, 125–26*t*, 130*t*
 population of, 125*t*
 size of firms in, 141
Malaysia
 analytical techniques used to assess economy of, 14–15
 automotive parts industry in, 59*t*, 60, 106*t*, 109–13, 163–65, 164–65*t*
 business groups in, 12–13, 13*n*24
 competitiveness and openness in, 14

Malaysia *(Continued)*
  electronics industry in, 105–9. *See also* electronics industry
  evolution of economy in. *See* evolution of Malaysian economy
  exports from, 37–92. *See also* exports from Malaysia
  FDI in, 98–104. *See also* foreign direct investment
  GDP growth rates in, 3n6, 18t, 21n8, 30, 189t, 191f, 197f. *See also* gross domestic product
  growth prospects for, 187–201. *See also* growth prospects
  imports into, 93–97, 94f, 96f
  innovation in, 119–86. *See also* innovation in Malaysia
  IRCA and RCA in, 210–11t
  macroeconomic indicators in, 28–30, 29t
  manufacturing sector, 22–30. *See also* industrialization and manufacturing
  MNCs, attracting, 7
  natural resources in, 2, 4, 5t, 8
  palm oil sector, 6n11, 58, 106t, 113–17. *See also* Malaysian Palm Oil Board
  physical infrastructure of, 30
  policy recommendations for, 203–18. *See also* policy recommendations
  political stability of, 6, 7
  population of, 6, 16
  poverty levels in, 16
  reasons for focusing on, 15–16
  technology transfers, 95–97, 96f
  as tiger economy, 2–3
  wages in, 24, 26t
Malaysia Venture Capital Management (MAVCAP), 153, 154, 157
Malaysian Agricultural Research and Development Institute (MARDI), 165–67, 166t, 168t
Malaysian Industrial Development Authority (MIDA), 17
Malaysian Institute of Microelectronic Systems (MIMOS), 107, 138, 165, 166t, 168t
Malaysian Palm Oil Board (MPOB), 116n27
  patenting activities, 172, 174, 176t
  R&D spending, 165–67, 166t, 168t, 176t
Malaysian Venture, 154
manufacturing. *See* industrialization and manufacturing
mapping product spaces, 81–86, 83–85f, 105, 112, 225–33t

MARDI (Malaysian Agricultural Research and Development Institute), 165–67, 166t, 168t
Mauritius, 89
MAVCAP (Malaysia Venture Capital Management), 153, 154, 157
medicine
  biomedical sector, 106n1, 181–82
  medical tourism, 89–90, 105n1, 210
megacities (agglomeration economies), 119–26, 123–30t, 216–17. *See also* specific Malaysian cities
mergers and acquisitions, 155–56
Mexico, 97n8, 160, 161f, 163f, 175f
microelectromechanical systems, 107
MIDA (Malaysian Industrial Development Authority), 17
middle-income trap, 3, 26–28
MIMOS (Malaysian Institute of Microelectronic Systems), 107, 138, 165, 166t, 168t
MNCs. *See* multinational corporations
Motorola in Malaysia, 172, 173t, 176t, 181
MPOB. *See* Malaysian Palm Oil Board
Multimedia Super Corridor (MSC), Malaysia, 149–50
multinational corporations (MNCs)
  development of Southeast Asian tiger economies, role in, 6
  in electronics industry, 107, 181
  export diversification, role in, 62, 66–67
  FDI from, 101–4
  innovation and diversification capacities and, 10–11
  Malaysian encouragement of, 7, 203, 205t, 209, 214
  MSC, investment in, 150
  parts and components sourcing by, 61, 77
  patenting activities, 172, 176
  shift from Singapore to China, 43
  technology transfers, 95, 97n7

N
National Venture Capital Promotion Fund, China, 157
natural resources as basis for development, 2, 4, 5t
NCER (Northern Corridor Economic Region), Malaysia, 217n11
neighboring economies, effects of, 199–201
NEP (New Economic Policy), Malaysia, 11, 12–13, 17

Netherlands, 123n15, 147t
New Economic Policy (NEP), Malaysia, 11, 12–13, 17
Northern Corridor Economic Region (NCER), Malaysia, 217n11
Norway, 212

O

OECD. *See* Organisation for Economic Co-operation and Development
oil palm cultivation and products in Malaysia, 6n11, 58, 106t, 113–17. *See also* Malaysian Palm Oil Board
oligopoly, problem of, 12–14
openness and competitiveness, 14. *See also* competition
Optometrix, 107
Organisation for Economic Co-operation and Development (OECD)
    as business environment, 34
    education in, 194f, 196f
    GDP growth in, 187–88, 188–89f, 189t
    IT infrastructure in, 144–46
    R&D spending in, 191–92, 193t

P

Pakistan, 2
palm oil sector in Malaysia, 6n11, 58, 106t, 113–17. *See also* Malaysian Palm Oil Board
patenting activity, 172–74t, 172–78, 175f, 176–78t
Penang, Malaysia
    education and knowledge economy in, 138–39t
    electronics industry in, 17n1, 106–9, 125n17, 181
    FDI flows in, 103, 104, 126t
    as free trade zone, 7
    GDP of, 125t
    industrialization and changing export mix in, 8
    as innovative urban location, 124–26, 125–26t, 129t
    population of, 125t
    service economy of, 87, 89–91, 210
    size of firms in, 140, 141
Penang Skills Development Centre (PSDC), Malaysia, 102n14, 108–9, 134
Pentamaster, 140, 182–83

Petronas, 140, 162, 181
Philippines
    automotive industry in, 110
    business environment, 34t
    in competitiveness indexes, 31–33t
    development path of, 6, 7
    electronics industry in, 107
    exports from
        competitive overlap, 47, 49
        diversification, 64–65t, 67, 68t
        electronics market share, 43f
        export structure, 9t
        growth rates, 39t
        industrial change detected from, 45t
        mapping product spaces, 82, 84f, 86, 231t
        quality of products, 76–77t, 81n21
        specialization, 61–62f
    FDI to, 98
    GCCs in, 12
    growth rates for
        education, 193–99, 194f, 198f
        GDP, 30, 189t, 191, 192f, 198f
    innovation in
        access to finance, 151f, 152f
        educational attainment, 134, 135–37t, 135f
        IRCA, 238t
        IT infrastructure, 145t, 146, 147t, 148f
        licensing and technology transfers, 179–80f
        patenting activities, 172, 174t, 175f, 177–79t
        R&D spending, 161–64f, 161t
    macroeconomic indicators, 29t, 30
    natural resources in, 2, 4, 5t, 8
    policy recommendations for, 216, 217n12
    RCA, 221t
    service economies, 87, 91, 92
    technological capabilities, 26, 28t
    tiger economy of, 2–3
photonics, 107
Plaza Accord (1985), 8n19, 67n17
policy recommendations, 203–18
    education and knowledge economy, fostering, 203–8, 204t
    industrialization and manufacturing, 208–14, 210–13t
    IRCA and RCA, 209, 210–13t
    IT infrastructure, 216
    macroeconomic policy, 217–18
    for private firms, 211–16
    productivity issues, 209, 216

policy recommendations *(Continued)*
  technology incentives, 203–8, 204–7t, 209–16
  urban agglomeration economy, need for, 216–17
political authoritarianism, 6, 7
polycentric urban systems, 123
poverty levels in Malaysia, 16
private firms
  large firms and innovation, 139–44, 142f, 143–44t, 214–16
  policy recommendations for, 211–16
  R&D activities
    by industry sector, 163–65, 164–65f, 166t
    large firms, 139–41, 143t
    in Malaysia, 180–83, 235–36t
    SMEs, 109, 111, 112, 211–14
private investment, role of, 18–21
product space analysis, 81–86, 83–85f, 105, 112, 225–33t
productivity issues
  in agglomeration economies and clusters, 120n3–5
  automotive industry, 111
  exports, PRODY value of, 74–77, 76–77t
  IT infrastructure, 147, 148f
  manufacturing sector, 23–30, 25–28t
  neighboring economies' effects on Malaysia, 199–200
  policy recommendations for, 209, 216
  TFP growth, 19–22, 20–21t, 28–30, 106
  value added per worker, 23–25, 25t, 27f–t
Promotion of Incentives Act of 1986 (Malaysia), 17
Proton Holdings Berhad, 109–10, 112, 164–65
PSDC (Penang Skills Development Centre), Malaysia, 102n14, 108–9, 134
public investment as growth driver, 18t, 19
Putrajaya, Malaysia, 138t, 141, 144t, 149n35

Q

quality
  effects of neighboring economies on, 200–201
  of tradable products, 69–77, 70–73f, 76–78t

R

Ramos, Fidel V., 7
RAND, 28t
R&D. *See* research and development
Razak, Najib, 217–18
Razak, Tun Abdul, 217

RCA (revealed comparative advantage), 77–81, 79–81t, 209, 210–13t, 219–23t
receiver-active approach, 159n1
research and development (R&D)
  as analytical tool, 15
  in automotive industry, 111
  capacity for innovation and, 10, 22
  in electronics industry, 108, 165f, 167t
  FDI in, 104
  on global competitiveness indexes, 32
  in palm oil industry, 117
  policy incentives to encourage, 203–8, 204–7t, 209–16
  private firms engaging in
    by industry sector, 163–65, 164–65f, 166t
    large firms, 139–41, 143t
    in Malaysia, 180–83, 235–36t
  in rubber industry, 113
  spending on, 159–71
    comparative long-term growth statistics, 191–92, 193t
    concentration by field, 170–71, 170t
    GDP, as share of, 160t, 161f, 191–92, 193t
    by GRIs, 165–67, 167–68t, 171t–f, 185
    personnel engaged in R&D, 160–62, 161–63f, 161t
    by private industry sector, 163–65, 164–65f, 166t, 180–83
    by universities, 167–70, 169t, 171t–f
  technology transfers, 95–97, 96f
  at tertiary institutes, 137, 138–39t, 167–70, 169t, 171t–f
  in urban agglomeration economies, 120
revealed comparative advantage (RCA), 77–81, 79–81t, 209, 210–13t, 219–23t
royalties and licensing, 178–80, 179–80f
rubber industry in Malaysia, 8n18, 15, 17, 22, 24, 42n6, 53, 113
Russian Federation, 124

S

San Diego, cluster development in, 103, 140, 142f
savings as share of GDP, 28–30, 29t
Second Science and Technology Policy (STEP2), Malaysia, 160n3, 205t
Selangor, Malaysia, 109, 113, 141, 144t, 167
service sector
  in Malaysia
    exported services, scope for expanding, 87–92, 88–89f, 90t
    sectoral contribution to growth, 19t

policy recommendations for, 208–10
in Southeast Asia compared to other countries, 10
Silicon Valley, 103, 124, 140, 150, 157
Sime Darby, 113, 116
Singapore
  as active FDI-dependent learning country, 159n1
  business environment, 34t
  business groups in, 13n24
  in competitiveness indexes, 31–33t, 34
  education growth rates in, 194f
  electronics industry in, 107
  exports from
    competitive overlap, 47
    diversification, 64–65t
    electronics market share, 43f
    mapping product spaces, 86, 232t
    MNCs as source of, 102
    as percentage of GDP, 37, 38t
    plants and equipment imported to Malaysia, 96
    quality of products, 75, 77t
    RCA, 80
  FDI from, 101f
  FDI to, 98
  import tariffs, 93
  innovation in
    access to finance, 151–52f, 154f, 155t
    educational attainment, 134, 135–37t, 135f
    index of innovation comparative advantage, 184, 185t
    IRCA, 239t
    IT infrastructure, 145t, 147t, 149t, 150
    licensing and technology transfer, 179–80f
    patenting activities, 173, 174t, 175f, 176–78t, 178
    R&D spending, 160t, 161–64f, 161t
    urban agglomeration economy, 123–24t
  investment effects on Malaysia, 200
  macroeconomic indicators, 29t
  Malaysia compared, 7
  policy recommendations drawn from, 217
  RCA, 222t
  service economies in, 10, 89, 90n26, 91–92
  technological capabilities in, 26, 28t
  tiger economy of, 1–2
  value-added ratios in machinery, 27f

SIRIM (Standards and Industrial Research Institute of Malaysia), 138, 165–67, 166t, 168t, 173t
skilled workforce. *See* education and knowledge economy
Small and Medium Industry Development Corporation, 109
small and medium-size enterprises (SMEs), 109, 111, 112, 211–14
SMEs (small and medium-size enterprises), 109, 111, 112, 211–14
South Africa, export of Malaysian cars to, 112n11
Southeast Asian tiger economies, 2–3. *See also* Indonesia; Malaysia; Philippines; Thailand
  analytical techniques, 14–15
  development history of, 4–7, 5t
  East Asian model followed by, 5
  exports from, 37–92. *See also* exports from Malaysia, and under other specific countries
  FDI, 98–104. *See also* foreign direct investment
  growth prospects, 187–201. *See also* growth prospects
  imports, 93–97, 94f, 96f
  industrial/export structure, 8, 9t
  industrialization and manufacturing in. *See* industrialization and manufacturing
  innovation in, 119–86. *See also* innovation in Malaysia, and under other specific countries
  macroeconomic indicators, 29t
  openness and competitiveness in, 14
  policy recommendations for, 203–18. *See also* policy recommendations
  vulnerabilities felt by, 3, 14, 22
S.P. Radio (Denmark), 141
specialization, role of, 60–62, 61–62f, 208–9
Standards and Industrial Research Institute of Malaysia (SIRIM), 138, 165–67, 166t, 168t, 173t
STEM (science, technology, engineering, and math) skills, importance of emphasizing, 11
STEP2 (Second Science and Technology Policy), Malaysia, 160n3, 205t
Straits Orthopaedics, 181–82
sub-Saharan Africa, 30. *See also* specific countries
Suharto, 6
Supermax, 113
Sweden, 122n12, 146–47t
Switzerland, 146–47t, 163, 165f, 208

T
Tai Kwong Yokohama, 181
Taiwan, China
  automotive industry in, 110
  backward links from MNCs in, 10
  business environment, 34*t*
  in competitiveness indexes, 31–33*t*
  electronics industry in, 107
  export-led growth in, 5
  exports from, 39*t*, 43*f*, 45*t*, 47, 49, 64–65*t*, 65, 75, 77
  FDI from, 101*f*
  growth rates for
    education, 194*f*, 196*f*
    GDP, 189*t*, 192*f*, 196*f*
    R&D spending, 192, 193*t*
  innovation in
    access to finance, 150, 151, 157
    educational attainment, 134, 135–36*t*
    IT infrastructure, 145*t*, 147, 149*t*
    large firms, 139, 140*n*27
    patenting activities, 173, 174*t*, 175*f*, 177–78*t*, 178
      R&D spending, 161*f*, 163*f*
  macroeconomic indicators, 29*t*
  Malaysia compared, 10, 11
  natural resources in, 4
  policy recommendations drawn from, 208, 217
  RCA, 222*t*
  technological capabilities in, 26
  tiger economy of, 1–2
tariffs on imports to Malaysia, 93, 94*f*
technological capabilities, 25–28, 28*t*, 32–34, 33*t*
technology clusters, 103, 106, 120, 124, 140–41, 142*f*
technology incentives, 203–8, 204–7*t*, 209–16
technology spillovers from FDI, 102–4
Technology Transfer from Research Institutes to SMEs (TEFT), Norway, 212–13
technology transfers, 95–97, 96*f*, 178–80
TEFT (Technology Transfer from Research Institutes to SMEs), Norway, 212–13
Tekes (Finland), 212
TFP (total factor productivity) growth, 19–22, 20–21*t*, 28–30, 106
Thailand
  automotive industry in, 110, 111, 112
  business environment, 34*t*
  business groups in, 13*n*24
  in competitiveness indexes, 31–33*t*, 32, 34
  development path of, 6
  electronics industry in, 107
  export of Malaysian cars to, 112*n*11
  exports from
    competitive overlap between Malaysia and, 47
    diversification, 64–65*t*, 66–67, 68*t*
    electronics market share, 43*f*
    exports from, 75
    growth rates, 39*t*
    industrial change detected from, 45*t*
    mapping product spaces, 82, 84*f*, 86, 233*t*
    MNCs as source of, 102
    plants and equipment imported to Malaysia, 96
    quality of, 71, 73, 77*t*, 81*n*21
    specialization, 61, 62*f*
    structure of, 9*t*
  FDI to, 98
  GCCs in, 12
  growth rates for
    education, 194*f*, 198*f*
    GDP, 189*t*, 193*f*, 198*f*
  innovation in
    financial incentives for, 245–48*t*
    IRCA, 212–13*t*, 239*t*
    licensing and technology transfers, 179–80*f*
    patenting activities, 172, 174*t*, 175*f*, 177*t*, 178
    R&D spending, 160–61*t*, 161–64*f*, 163
  investment effects on Malaysia, 200
  IRCA, 212–13*t*, 239*t*
  macroeconomic indicators, 29*t*, 30
  medical tourism, 89
  natural resources in, 2, 4, 5*t*, 8
  Plaza Accord (1985) affecting economy of, 67*n*17
  policy recommendations for, 203, 209, 212–13*t*, 216
  RCA, 80, 212–13*t*, 223*t*
  service economies, 87, 89, 90*n*27, 92
  technological capabilities in, 26, 28*t*
  TFP growth in, 20, 21*t*
  tiger economy of, 2–3
  urban agglomeration economies in, 123–24*t*
tiger economies, 1–16
  cultural factors associated with, 2
  in East Asia. *See* East Asian tiger economies; Hong Kong, China; Korea, Republic of; Singapore; Taiwan, China
  first tiger economy in Japan, 1
  in Southeast Asia. *See* Indonesia; Malaysia; Philippines; Southeast Asian tiger economies; Thailand

Top Glove, 113
total factor productivity (TFP) growth, 19–22, 20–21*t*, 28–30, 106
tourism, 87
trade
 exports. *See* exports from Malaysia
 imports, 93–97, 94*f*, 96*f*
 services. *See* service sector
 specialization in tradable products and services, 208–9
trade policies assisting tiger economies, 5–6

U

UN Comtrade (United Nations Commodity Trade Statistics Database), 69, 73
UNDP (United Nations Development Programme), 28*t*
unit values, 69–74, 70–73*f*
United Kingdom
 export of Malaysian cars to, 112*n*11
 growth rates for
  education, 196*f*
  GDP, 187, 188*f*, 196*f*
  R&D spending, 191, 193*t*
 IT infrastructure in, 146–47*t*
 large firms and cluster development in, 103, 140–41, 144
United Nations Commodity Trade Statistics Database (UN Comtrade), 69, 73
United Nations Development Programme (UNDP), 28*t*
United States
 economic and policy assistance for developing economies, 5–6
 FDI from, 1, 101*f*
 GDP growth rates for, 187, 189*f*
 imports into, 37, 39, 40*f*, 51–54, 52*t*, 55*t*, 58*t*, 77

innovation in
 access to finance, 150, 157
 educational attainment, 135–36*t*
 IT infrastructure, 145, 146–47*t*
 large firms, 140–41, 142*f*
 patenting activities, 177–78*t*
 R&D spending by sector, 165*f*
 urban economies, 123*n*16, 124, 126*n*18
MNCs from, 6, 7
productivity in, 19, 21*n*7
R&D investment in, 95
technology transfers to Malaysia from, 96
universities. *See* education and knowledge economy
urban agglomeration economies, 119–26, 123–30*t*, 216–17. *See also* specific Malaysian cities

V

value added per worker, as measure of productivity, 23–25, 25*t*, 27*f*–*t*
venture capital, 151–57, 153*t*, 154*f*, 155*t*
Vietnam, 108

W

wages in Malaysia, 24, 26*t*
Washington Consensus, 7
Weber, Max, 2*n*4
WEF (World Economic Forum), 26, 28*t*, 32*t*
workforce. *See* education and knowledge economy; labor and employment issues
World Bank, 15, 26, 34, 49*n*10, 150, 203, 207
World Economic Forum (WEF), 26, 28*t*, 32*t*
World Trade Organization (WTO), 49, 110
worldwide recession of 2008–2009, 187
WTO (World Trade Organization), 49, 110

Y

Yozma Program, Israel, 157

ECO-AUDIT
Environmental Benefits Statement

The World Bank is committed to preserving endangered forests and natural resources. The Office of the Publisher has chosen to print *Tiger Economies Under Threat* on recycled paper with 30 percent post-consumer waste, in accordance with the recommended standards for paper usage set by the Green Press Initiative, a nonprofit program supporting publishers in using fiber that is not sourced from endangered forests. For more information, visit www.greenpressinitiative.org.

Saved:
- 5 trees
- 129 lb. of solid waste
- 2,132 gal. of waste water
- 443 lb. of net greenhouse gases
- 1 million Btu of total energy